PERSONAL NETWORKS

PERSONAL NETWORKS
WIRELESS NETWORKING FOR PERSONAL DEVICES

Martin Jacobsson
Delft University of Technology, The Netherlands

Ignas Niemegeers
Delft University of Technology, The Netherlands

Sonia Heemstra de Groot
Delft University of Technology, The Netherlands and Twente Institute of Wireless and Mobile Communications, The Netherlands

WILEY

A John Wiley and Sons, Ltd., Publication

Library of Congress Cataloging-in-Publication Data

Jacobsson, Martin, 1976-
  Personal networks : wireless networking for personal devices / Martin Jacobsson, Ignas Niemegeers, Sonia Heemstra de Groot.
      p. cm.
  Includes bibliographical references and index.
  ISBN 978-0-470-68173-2 (cloth)
  1. Wireless communication systems.   2. Personal communication service systems.   3. Ubiquitous computing.   I. Niemegeers, Ignas.   II. Heemstra de Groot, Sonia.   III. Title.
  TK5103.2J34 2010
  621.384 – dc22

                                                                        2010005593

A catalogue record for this book is available from the British Library.

ISBN 978-0-470-68173-2 (H/B)

Set in 10/12 Times by Laserwords Private Limited, Chennai, India
Printed and Bound in Singapore by Markono Print Media Pte Ltd.

Contents

Foreword

The personal network (PN) vision is essentially that people's access to digital assets (all the devices that they own and their contents) should be made simple and convenient at any time and from any location. As with any vision, this is easily enough stated yet extremely difficult to realize fully. There is certainly much more to it, in terms of technical challenges and potential benefits, than the vision would seem to imply.

Some readers may not see much distinction between the PN ambition and what is readily available today in a smart phone. Others may understand that the PN is beyond current capabilities but may not see why anyone would want to adopt it. Yet others may find the notion of PNs desirable but believe that it is really unattainable.

This book anticipates the questions raised by each of the above viewpoints. It presents visions in the form of future scenarios, and the associated future user requirements in more technical terms. The current know-how in personal networking and where it is going next are also covered. These early chapters should provide the uninformed or skeptical reader with the necessary incentive to read further. They also convey the tremendously exciting possibilities offered by PNs across various walks of life.

The bulk of the book is about how PNs might be realized, starting with a description of the architecture in which the necessary technical elements would be combined. Each of the main technical issues is covered in detail in separate chapters that show how the user's access to digital assets can be achieved – MANET clusters, routing and tunneling between clusters, communication with so-called 'foreign devices', applications support and security implications. Three prototype personal network systems are outlined, including the authors' own at the Delft University of Technology. Finally, there is a brief look ahead exploring what PNs may be like in the future.

This welcome new volume in the Wiley Series in Communications Networking & Distributed Systems is written by three of the leading experts who have been immersed for the past several years in the challenge of building personal networks. It gives a comprehensive and distinctive coverage of this important field and should appeal broadly to researchers and practitioners in the field of communications and computer networks as well as to those specifically enthused by the prospect of personal networking.

David Hutchison
Lancaster University

Preface

Recent decades have shown a tremendous expansion of the Internet. The number of connected terminals has increased by orders of magnitude, traffic has grown exponentially, coverage has become ubiquitous and worldwide, and today's sophisticated Web 2.0 applications are increasingly providing services which hitherto have been the realm of telecommunications, such as Skype and video conferencing. This has even led to the thought that access to the Internet might one day be a universal right of every citizen. This evolution will accelerate in the coming decades. The driving factor is mobile Internet, a result of the continuing validity of Moore's law, according to which the density of microelectronic circuitry doubles every year and a half. The implication is that computing power and, in its wake, communication power will continue to increase exponentially. Its corollary is a fall in the cost of providing a certain amount of computing and communication power to the extent that it is becoming perfectly feasible to equip every artifact with computing and communication capabilities. This is what enables 'the Internet of things' – it is expected that there will be of the order of 1000 devices per person in the year 2017 (Tafazolli 2004). The range of device types and their capabilities will be mind-boggling. Most of these devices will be mobile or at least wirelessly connected. A huge challenge will be to exploit this sea of devices and their connectedness to create novel and useful applications without drowning in the complexity of managing large heterogeneous distributed systems.

The vision of personal networks was based on these trends, which were foreseeable given Moore's law and the derived technology roadmaps. It was the result of brainstorming sessions taking place in 2000 at Ericsson Research and Delft University of Technology in the Netherlands. The dream was to create an environment in which every person has at his fingertips all the digital devices he owns regardless of where he or she is and where those devices are, as long as they are connected. We envisaged a person to be always surrounded by a 'virtual digital bubble' formed by his or her personal devices. This personal network would enhance a person's private and professional capabilities in terms of access to information, control of his environment, social interaction, etc. It would dynamically change as the person moved around and engaged in different activities. It would have a global reach and would always incorporate those devices that are most suitable to support the person.

As we began exploring the idea, we came upon the Moped project of Robin Kravetz at the University of Illinois, which had a similar vision. This together with other ideas triggered the concept of a personal network, the subject of this book. The ideas were elaborated in two large European research projects, MAGNET and MAGNET Beyond, and two Dutch projects, PNP2008 and QoS for PN@Home. These led not only to the development and prototypes of technical solutions for the basic functionalities required in personal networks, but also to first experiments with applications and the study of potential business models. The concrete solutions that are presented in this book were developed in those projects. In parallel, similar ideas had been developed in the UK in the context of the Mobile Virtual Centre of Excellence by James Irvine and John Dunlop at the University of Strathclyde. Their concept was named the 'personal distributed environment'.

The ideas about personal networks were very much centered on the person and how her capabilities could be enhanced by creating a synergetic environment consisting of the hundreds of personal digital devices she might own in the near future. A natural next step was to explore how similar synergies could be achieved by pooling personal resources to support and enhance the activities of a group of people. This led to the concept of federations of personal networks belonging to different people. These ideas were also explored in the MAGNET Beyond and PNP2008 projects.

The basic foundations have been laid to build personal networks and their federations, and prototypes and demonstrators have been built. More research is needed in particular to create environments that allow rapid development of personal network applications, and to facilitate different business roles and models to make the concepts commercially viable. This will also require efforts in standardization, which have already started.

The market pull to build and use personal networks and their federations is not there yet. However, we believe that we are on the brink of a breakthrough in this respect. If the WWRF predictions of 1000 devices per person in 2017 and the 100 billion mobile Internet devices in the next decade foreseen by Cisco (Cisco Systems 2009) become a reality, concepts such as personal networks will be good tools not only to manage the resulting complexity, but, even more importantly, to create hitherto unknown opportunities to empower people in their private lives and at work.

This book covers the core concepts of personal networks and federations of personal networks, and explains their architecture. It elaborates in detail the various aspects of these architectures, including topics such as networking, self-configuration, security, personal services, service sharing, and context management. It also discusses the outcomes of several personal network research projects, including the prototypes. It is aimed at researchers, developers, and standardization experts in mobile and wireless communication systems and services. It should also be of interest to graduate students in the field of telecommunications and distributed systems.

The book is organized as follows. The introductory Chapter 1 describes the vision underlying personal networks. This is followed by Chapter 2 that set the stage by discussing user requirements and Chapter 3 that covers trends in personal networking. Readers who are up to date on the state of the art of developments in wireless and mobile technologies and applications and ubiquitous computing may go straight to Chapter 4 that discusses the personal network architecture. Next are several detailed chapters that may be read independently: Chapter 5 on cluster formation and routing, Chapter 6 on inter-cluster tunneling, Chapter 7 on communication between a personal network and entities that do not belong to it (the so-called foreign devices), Chapter 8 on application support, Chapter 9 on security, and Chapter 10 on federations of personal networks. Chapter 11 introduces three different existing personal network prototypes that builds on the concepts introduced in the previous chapters. The book is rounded off by Chapter 12, which gives the authors' view on the future of personal networks.

Acknowledgments

We have already acknowledged the projects that led to the elaboration of the personal networks and their federations, but we should in particular acknowledge the hard work of many PhD and MSc students at Delft University of Technology and other universities across Europe who contributed research results and building blocks. We should also mention leading research institutes, such as the University of Cantabria, IBBT, Télécom & Management Paris Sud, TNO, VTT, LETI, and CSEM and companies, such as TI-WMC, NEC, Nokia, Telia Sonera, Philips, and KPN that played a big role in the projects we mentioned. Furthermore, we gratefully acknowledge the work of our colleagues at TI-WMC and Delft University of Technology involved in the research and development of personal networks.

We must mention three persons who believe in our ideas and have given us strong support: John de Waal of Ericsson Research and co-founder of TI-WMC, through interactions with whom the concept of personal networks took shape; Dr. Jorge Pereira of the European Commission who saw the potential and challenged and encouraged us; and Prof. Ramjee Prasad of Aalborg University who carried the heavy load of managing the MAGNET and MAGNET Beyond projects.

We are also grateful to Jereon Hoebeke (IBBT) and Kimmo Ahola (VTT) for providing us with screen shots of the MAGNET prototype. Last but not least, we would also like to thank the people, including Sabih Gerez and Torsten Jacobsson, who read earlier versions of this book and provided valuable feedback.

Martin Jacobsson, Ignas Niemegeers, Sonia Heemstra de Groot
Delft, The Netherlands

List of Abbreviations

3GPP	Third Generation Partnership Project
AAA	Authentication, Authorization, and Accounting
ADSL	Asymmetric Digital Subscriber Line
AIPN	All-IP Networks
AN	Ambient Networks
AODV	Ad Hoc On-Demand Distance Vector
API	Application Programming Interface
BAN	Body Area Network
CA	Certification Authority
CAN	Community Area Network
CBB	Counter-Based Broadcasting
CMI	Context Management Interface
CMN	Context Management Node
CoA	Care-of Address
CPFP	Certified PN Formation Protocol
CRL	Certificate Revocation List
CS	Certificate Server
CTS	Clear to Send
DA	Directory Agent
DAD	Duplicate Address Detection
DB	Database
DCF	Distributed Coordination Function
DHCP	Dynamic Host Configuration Protocol
DHT	Distributed Hash Table
DME	Device Management Entity
DNA	Detecting Network Access
DNS	Domain Name System
DoS	Denial-of-Service
DSDV	Destination-Sequenced Distance-Vector Routing
DSL	Digital Subscriber Line
DSR	Dynamic Source Routing
DYMO	Dynamic MANET On-Demand Routing Protocol
ECC	Elliptic Curve Cryptography
ER	Edge Router
ESP	Encapsulating Security Payload
ETT	Expected Transmission Time

ETX	Expected Transmission Count
EWMA	Exponentially Weighted Moving Average
FA	Foreign Agent
FIFO	First In First Out
FMIPv6	Fast Handover for Mobile IPv6
FP6	Sixth Framework Programme
FSP	Flooding with Self-Pruning
GENA	General Event Notification Architecture
GLL	Generic Link Layer
GPS	Global Positioning System
GSM	Global System for Mobile Communication
HA	Home Agent
HDMI	High Definition Multimedia Interface
HI	Host Identifier
HIP	Host Identity Protocol
HMIPv6	Hierarchical Mobile IPv6
i3	Internet Indirection Infrastructure
ICMP	Internet Control Message Protocol
ICT	Information and Communication Technology
IEEE	Institute of Electrical and Electronic Engineers
IETF	Internet Engineering Task Force
IKE	Internet Key Exchange
INR	Intentional Name Resolver
INS	Intentional Naming System
IP	Internet Protocol
IPC	Inter-Process Communication
IPsec	Internet Protocol Security
IrDA	Infrared Data Association
ISM	Industrial, Scientific, and Medical
ISTAG	Information Society Technologies Advisory Group
IST	Information Society Technology
LLAL	Link Layer Adaptation Layer
LLC	Location Limited Channel
LoC	Lines of Code
LQA	Link Quality Assessment
LTE	Long Term Evolution
MAC	Medium Access Control
MAC	Message Authentication Code (Chapter 9)
MAGNET	My Adaptive Global Net
MANET	Mobile Ad Hoc Network
MIH	Media Independent Handover
MMS	Multimedia Messaging Service
MOPED	Mobile Grouped Device
MPR	Multipoint Relay
MR	Mobile Router
MSMP	MAGNET Service Management Platform
MTM	Medium Time Metric
MTU	Maximum Transmission Unit
N3	Notation 3

NAPT	Network Address Port Translation
NAT	Network Address Translator
NEMO	Network Mobility
NEXWAY	Network of Excellence in Wireless Applications and Technology
NFC	Near Field Communication
NHDP	Neighborhood Discovery Protocol
OLSR	Optimized Link State Routing Protocol
P2P	Peer-to-Peer
PAC	Proximity Authenticated Channel
PACWOMAN	Power Aware Communications for Wireless Optimised Personal Area Network
PAN	Personal Area Network
PC	Personal Computer
PDA	Personal Digital Assistant
PDE	Personal Distributed Environment
PFS	Prioritized Flooding with Self-Pruning
PGP	Pretty Good Privacy
PKI	Public Key Infrastructure
PMH	Personal Mobile Hub
PN	Personal Network
PNCA	PN Certification Authority
PNDB	Personal Network Database
PNDS	PN Directory Service
PNF	PN Federation
PNNT	Personal Node Neighbor Table
PNP2008	Personal Network Pilot 2008
PNPA	PN Provisioning Administration
P-PAN	Private Personal Area Network
PRNET	Packet Radio Network
PVR	Personal Video Recorder
QoS	Quality of Service
ROAM	Robust Overlay Architecture for Mobility
RSSI	Received Signal Strength Indication
RTS	Request to Send
RVS	Rendezvous Server
SCMF	Secure Context Management Framework
SDN	Service Directory Node
SHAMAN	Security for Heterogeneous Access in Mobile Applications and Networks
SIM	Subscriber Identity Module
SLP	Service Location Protocol
SMN	Service Management Node
SMS	Short Message Service
SNR	Signal to Noise Ratio
SOAP	Simple Object Access Protocol
SPI	Security Parameter Index
SSDP	Simple Service Discover Protocol
STUN	Session Traversal Utilities for NAT
TCP	Transmission Control Protocol
TEP	Tunnel Endpoint

TLS	Transport Layer Security
TTP	Trusted Third Party
TURN	Traversal using Relay NAT
UCL	Universal Convergence Layer
UDP	User Datagram Protocol
UIA	User Information Architecture
UIP	Unmanaged Internet Protocol
UML	Unified Modeling Language
UMTS	Universal Mobile Telecommunications System
UPN	Universal Personal Networking
UPnP	Universal Plug and Play
USB	Universal Serial Bus
UWB	Ultra-Wide band
VoIP	Voice over IP
VPN	Virtual Private Network
WAN	Wide Area Network
WCETT	Weighted Cumulative ETT
WebDAV	Web-Based Distributed Authoring and Versioning
WiMAX	Worldwide Interoperability for Microwave Access, Inc.
WLAN	Wireless Local Area Network
WPA	Wireless Protect Access
WPAN	Wireless Personal Area Network
WRP	Wireless Routing Protocol
WSI	Wireless Strategic Initiative
WSN	Wireless Sensor Network
WWI	Wireless World Initiative
WWRF	Wireless World Research Forum
XACML	Extensible Access Control Markup Language
XML	Extensible Markup Language
X-RBAC	XML Role-Based Access Control

1

The Vision of Personal Networks

Since the dawn of time, communication has been an integral part of human life and the need for better technology to support our communication has been growing continuously. Over the centuries, we have invented many different methods of communication to bridge the barrier of distance. With people becoming increasingly nomadic, the need for communication with business partners all over the world and with loved ones at home while on the move has never been greater. This is the basis of the worldwide success of mobile telephony. Migrant workers overseas may easily, for a relatively small cost, have voice conversations with their family on the other side of the planet. At the same time, the mode of communication has become richer and more varied. Today, nothing stops us from sending video and audio messages to any place on earth.

1.1 Past, Present, and Future Telecommunication

Telecommunication technologies, both wired and wireless, are what make rich communication, such as voice or video, possible for people on the move. Information and communication technology (ICT), which is the merger of telecommunication and computing, is the major enabling factor. However, rich communication is not limited to human interaction. Technology is increasingly used to automate many tasks. For example, with home automation, we can, in principle, control every electronic device in our homes. With electronic agendas accessible from everywhere, we can better plan our daily activities. By using sophisticated entertainment devices, we can listen to music, watch movies, or play games while waiting at the bus stop or at the airport.

From its roots in ARPANET (Abbate 1999), the Internet started in 1969 as a research project and grew into a worldwide network in the second half of the 1990s, connecting computers all over the world. Popular services such as e-mail, the World Wide Web, peer-to-peer file sharing, and more recently social networking evolved and made the Internet attractive for private citizens, business, and government alike. The growth of the Internet has been remarkable, and it has reached 60% of the population in the Western world (http://www.internetworldstats.com/). But it does not stop there. While the rate of Internet penetration is slowing down, the achievable data rates continue to increase and

---

Personal Networks: Wireless Networking for Personal Devices   Martin Jacobsson, Ignas Niemegeers and Sonia Heemstra de Groot
© 2010 John Wiley & Sons, Ltd

this will enable new services. Soon it will be possible to broadcast television and video on demand over the Internet to everyone everywhere.

Mobile telephony is yet another example of a very successful technology (Dornan 2001). The first successful mass market deployment of mobile telephone systems started in the 1980s. In less than 20 years, the mobile phone has gone from being a rare and expensive device, accessible only to business people with an interest in high-tech gadgets, to a pervasive low-cost personal item for everybody. In many countries, mobile phones now outnumber landline telephones, with most adults and many children owning mobile phones. In 2008, there were 4.02 billion mobile subscribers worldwide but only 1.27 billion landline subscribers (https://www.cia.gov/library/publications/the-world-factbook/geos/xx.html). While the Global System for Mobile Communication (GSM) and the various forms of 3G networks, such as the Universal Mobile Telecommunication System (UMTS), are currently the leading mobile technology standards, others, such as Long Term Evolution (LTE), will soon take over. These technologies offer better packet switching support as well as higher data rates with similar support for mobility. Another recent promising technology that can bring high data rates to the mobile user is IEEE 802.16 (IEEE 2004b, 2006a), also known as WiMAX. With these technologies we will soon be able to watch movies while on the move. However, this is probably just the start of the hunt for higher data rates for mobile devices. Better battery technology or other miniaturized energy sources, and energy harvesting techniques, more computational power, and improved radio technology will undoubtedly offer better data rates, higher quality, and more communication possibilities, enabling a vast range of high quality mobile services.

While Internet and mobile telephony have been developed side by side, there is a growing trend to integrate the two. Nowadays, there are plenty of websites on the Internet where one can send Short Message Service (SMS) or Multimedia Messaging Service (MMS) messages to mobile phones. Conversely, we have mobile phones that can send e-mails and connect to the Internet. Beyond any doubt, this trend will continue as normal users do not wish to have separate networks, for example one when on the move and another one when at home. Instead, users expect the two networks to be fully integrated.

The evolution of radio communication has also given birth to another trend: medium and short range wireless communication. One of the first successful mass market products in this segment was the wireless local area network (WLAN) standard IEEE 802.11 (IEEE 1999) originally released in 1997. It was designed to make the LAN wires redundant in an office and was much more successful in this than any of its predecessors, such as the Infrared Data Association (IrDA) (http://www.irda.org/). When the enhanced version IEEE 802.11b came onto the market, its deployment really took off. So-called hotspots were installed where an IEEE 802.11b (and later IEEE 802.11g) access point could offer wireless Internet connectivity with data rates of several Mbps to devices, such as laptops and personal digital assistants (PDAs), within a range of up to about 100 meters. Millions of hotspots have been installed worldwide in strategic locations where people congregate and need to communicate. Examples are airports, train stations, coffee shops, hotels, and convention centers.

To connect wearable and handheld devices around a person, a range in the order of 10 meters is enough. This has led to the development of yet another branch of technologies that cover a wide range of data transmission rates, have low power consumption, but a limited range. They go under the term wireless personal area networks (WPANs) or just personal area networks (PANs), of which IEEE 802.15.1 (IEEE 2005) (commonly known as Bluetooth) is currently the most common technology. These technologies interconnect mobile phones, laptops, PDAs, sensors and other personal devices located within 10 meters in a seamless way with low enough power consumption for normal battery-powered devices. Typical WPAN communication takes place between a person's mobile devices, such as a camera requesting time and location information from a Global Positioning System (GPS) receiver to tag a picture or a mobile phone sending voice to a wireless headset. It can also support information sharing between two persons meeting on the street. For instance, they can share recently taken pictures or interesting locations (geographical data) one of them just visited. Even in this segment, very high data rate versions are to be expected in the near future, such as the IEEE 802.15.3 family (IEEE 2003, 2006b). For the more distant future, data rates in the order of Tbps are the new target for research projects.

Current research and development will bring us more specialized communication technologies that are optimized for a particular niche. Figure 1.1 shows the current landscape of wireless communication technologies. It shows how each technology targets a specific area. It is clear that the variety of technologies we will have to cope with is likely to increase. The downside to this trend is the multitude of radio interfaces and protocols, between which there is currently a clear lack of integration. The advent of software defined radio and cognitive radio will to a certain extent help to address this issue, by providing radios that, depending on application and context, adapt themselves.

The major challenge that remains is to build wireless distributed systems providing a wide spectrum of applications on top of a multitude of devices using highly heterogeneous radio communication technologies. We cannot expect the end-user to deal with this issue. Therefore, it is important to use these technologies in a complementary way and make them work together seamlessly.

Regrettably, very little effort has been made to integrate these different technologies. One rare example is the attempt to integrate WLAN and cellular technologies (Vulić 2009).

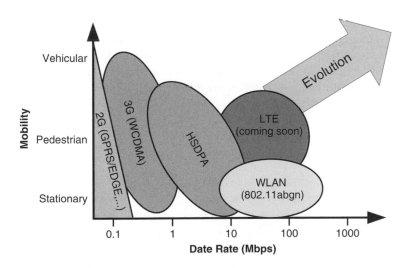

Figure 1.1 Wireless communication landscape.

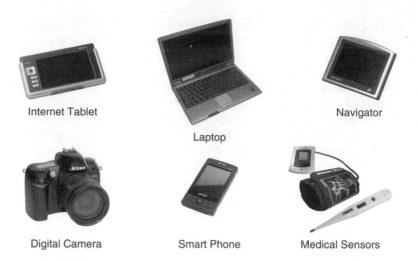

Internet Tablet   Navigator

Laptop

Digital Camera   Smart Phone   Medical Sensors

Figure 1.2   Examples of personal electronic devices.

Furthermore, it is possible to send e-mails from a mobile phone and SMSs from an Internet-connected personal computer (PC), but the possibilities should go well beyond this. Instead, users nowadays are forced to learn each system and manually configure it to interoperate. In many cases, this is simply not possible because of software limitations. This problem is bound to become worse as people make more and more use of electronic devices. At the same time, device technology has made rapid progress in recent decades. Our mobile phones become smart mobile computers and still retain their original form. Even the cheaper mobile phones of today can play music, take and display photos and video clips, and even surf the Web.

Driven by Moore's law (Moore 1965), the microelectronics industry has created ever smaller chips that are consuming less energy, are less costly, and yet are more powerful and capable of things one could hardly imagine before. This has led to a large variety of different devices and terminals, everything from small and simple mobile phones and music players to PDAs, tablet PCs, and computers embedded in virtually every artifact, to advanced mobile multimedia or entertainment platforms. Figure 1.2 shows some current examples. Hence, there is no reason why future terminals should be a limiting factor for enhanced interoperability.

1.2   Personal Networks

According to the Wireless World Research Forum (WWRF), by the year 2017, there will be 1000 wireless devices per person on earth (Jefferies 2007). These devices will vary from sophisticated multimedia systems to very simple sensor systems. Many of them will be intimately linked to people. They will be an important ingredient of what has been called 'the Internet of things' (Dodson 2003). In principle, this opens up the perspective of using this vast number of personal resources to enhance people's lives, professional and personal, regardless of where they are. However, the shortcomings of current wireless communication technologies are hampering the development of seamless communication between the multitude of devices a person will own. The careful reader will notice that most devices in Figure 1.2 have screens. These are needed because communication is cumbersome and forces us to interact directly with every single device, using screens and other input and output means.

In order to be successful, future information and communication technology should be centered on the user, improving the quality of life of and adapted to the individual, without the need for the user to be aware of the technical details. In order to achieve this, devices and environments need to become smarter, more responsive, and to accommodate the needs of the individual. Further, personalization and ubiquitous access to information and communication will be essential. Ideally, such a system must adapt to the situation and allow its users to use the most suitable means of communication and to access the most relevant information. As a consequence, new fields of research have emerged that aim to provide users with the same experience independent of user interfaces, terminal capabilities, communication technologies, and network and service providers. Examples of such fields are pervasive and ubiquitous computing (see Section 3.4) as well as ambient intelligence and ambient networking (see Section 3.5).

The personal network (PN) (Niemegeers and Heemstra de Groot 2003) is such a concept and technology. It is related to pervasive computing with a strong user-focused view. While a PAN connects a person's devices around her, a PN extends that PAN with other devices and services farther away. This extension will physically be made via any kind of wired or wireless network. This can include devices and networks around her in the car, office, or any other place. However, a PN is more than connectivity. A person's PN must support her applications and take into account her context, location and, of course, her communication possibilities. A PN must adapt to changes in the surroundings, be self-configuring and be able to incorporate many different types of networks and devices to be as useful as possible. Figure 1.3 shows what a PN could look like for a user. It shows how the user has electronic devices around her that can communicate with each other using WPAN technologies. It also shows how those devices can communicate with the devices of friends in the close vicinity as well as devices in smart buildings. The PN also incorporates devices elsewhere, such as in the office and at home.

There are many different ways of integrating the various communication technologies to achieve one unified system. The best and most complete integration approach is to define a common network layer to be used by all, which is similar to the approach taken by the Internet with the Internet Protocol (IP). Such a general and common network layer architecture that imposes minimal changes to the underlying network types, can bridge different communication technologies and offer a homogeneous and clear view to the end-user. At the same time, the network architecture needs to be future proof, that is able to accommodate all kinds of present and future applications and technologies. In order to be successful, a PN should cater for all of a person's communication needs. The PN must include not only the person's wearable and wireless devices but also devices at home, in the car, in the office, or any place where the user may have personal devices. This means that the network layer of the PN must work as a home network at home, a car network in the car, a PAN around a person and glue all these networks together in one PN. At the same time, it must cooperate with existing networks such as the Internet and other infrastructure networks.

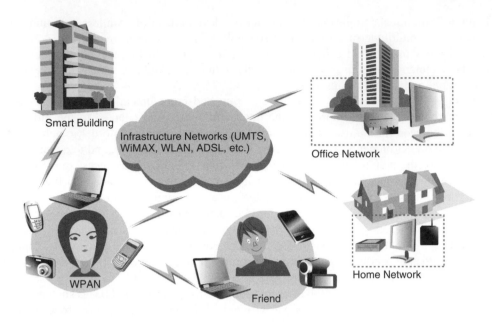

Figure 1.3   The concept of personal networks.

1.3   Some Typical PN Use-Case Scenarios

The success of PNs requires not only seamless integration at the network layer, but also, and more importantly, the development of new types of interesting and useful applications that exploit the full potential of PNs. To better introduce the concept of a PN, some use-case scenarios are given below that demonstrate the possibilities of a PN and what types of applications can benefit from a PN.

1.3.1   Introducing Jane

Let us meet Jane, who will pop up throughout this book to help us explain various PN concepts and how they apply to real users. When we refer to this example, we use indented and italic text.

Jane is a salesperson who travels a lot. For her, it is important to always be able to access her own data and services, regardless of their locations. Frequently, she needs to access information stored on computers in her office when she is on a company visit or on the way to the next meeting. To do this, Jane is equipped with a mobile phone, a laptop, a headset, and a navigation system.

Furthermore, Jane has a family with two children. To be away from home for extended periods of time can be demanding. However, screens, cameras, speakers, and microphones in her home enable her to have a richer form of communication with her family. The devices at home can provide her with a virtual home environment through which Jane can virtually see her family, talk to them, and even play games.

Sara is Jane's mother. Sara is aging, but still lives by herself, not far from Jane. However, Sara needs more and more attention, especially with household tasks such as cleaning and grocery shopping. Jane shares the task of helping her mother with her brother and one of Sara's neighbors. However, this requires a lot of coordination to, for instance, ascertain that Sara does not suddenly end up without food. To this end, Jane shares her agenda with the others so that better coordination can be achieved.

However, Jane has one problem with all this. There are so many applications, devices, and networks to keep track of and getting them to cooperate is a major task. Jane does not want to spend time on these sorts of issues and has therefore decided to create a PN for herself.

1.3.2 The Traveling Saleswoman

One major potential benefit of using PNs is seamless access to resources anywhere. For instance, personal files stored at home or in an office can be obtained by one's devices as long as there is network access. Figure 1.4 shows Jane during a company visit.

Jane's PN offers a framework that enables her devices to seamlessly cooperate and to communicate with distant devices, such as desktop computers, company servers, customer services, and home multimedia systems.

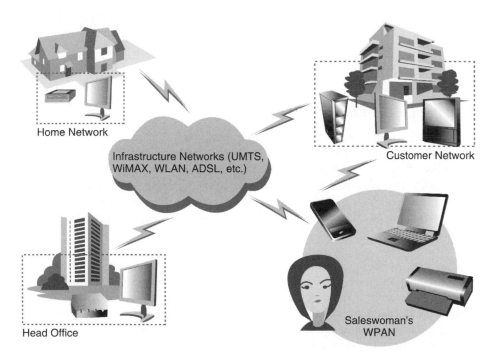

Figure 1.4 Traveling saleswoman scenario.

With a PN, Jane can easily access her agenda from any device wherever she is and at the same time make sure her secretary has an up-to-date copy as well. The same holds for personal and shared files. When at a client site, Jane can share some of these files with the client in order to be able to present products, make offers, etc. These are very simple applications, yet very important ones. They must work with whatever network access is available. For instance, when she is visiting a client, they should be able to use the client's network to improve transmission speed.

Furthermore, Jane's PN lets her communicate with her family using the equipment at home. The PN enables her to use the devices that she carries to communicate with the devices in her home and thereby offer her the ability to interact with her family in a rich way.

Depending on the communication requirements, she could also continue all this while traveling. She could listen to streamed music from the home multimedia system while driving, or play a game while waiting for an airplane, etc. If she meets a friend somewhere, a temporary network can be established, to share files, services or just to play a multi-player game for a while.

While several existing technologies can offer solutions to parts of this scenario, very little work has yet been done to combine these technologies into a seamless integrated solution for a normal user. Today, employers have experts who set up servers and configure wireless devices to interoperate with their enterprise software on behalf of their employees. Even so, these solutions are typically application-specific and will not work for new applications without proper integration. For the end-user, such as Jane, they are far from seamless. Complex settings cause frustrations and make people wonder whether it will work on the next customer visit. PNs try to address this issue by being easy to use, set up, configure, and maintain, as well as being fast and secure.

1.3.3   Care for the Elderly

PNs can be an even more powerful tool for personal communication if they are designed to interact with other PNs as well as existing networks and services. With an aging population, this may prove to be a very important function. An elderly person could be equipped with a PN consisting of various medical sensors to continuously allow monitoring of her health. Such sensors could include blood pressure and heartbeat sensors, activity sensors, accelerometers, and positioning devices. When something happens, the PN could alert any interested parties. Figure 1.5 illustrates this scenario.

Sara's doctor decides that it would be a good idea to monitor Sara more closely in case something happens and arranges for a wearable fall detector and some activity sensors to be placed in Sara's home. With PN technology, these sensors can trigger an alarm on some other predefined PNs. In this case, the system is configured to notify the PNs of Sara's daughter Jane, Sara's neighbor, and a special care organization. Using a camera in the home, any of these persons can try to make contact and find out more details when an incident occurs.

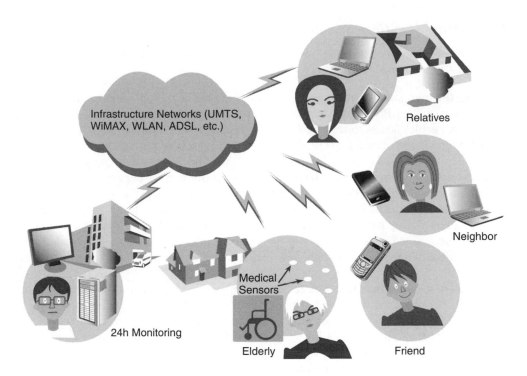

Figure 1.5   Care for an elderly person.

Sara is also offered a device that can trigger the alarm at the push of a button. That device can also track the location. When the button is pressed, the location can also be sent with the notification so that medical staff can be sent to the correct location immediately. Such a device may allow Sara to leave the house, knowing that help is still available if something happens.

A PN can also improve an elderly person's capability to communicate with friends, who might also be elderly, or it can remind them about various things, such as when to take certain medicines for those whose memory is fading. However, designing a PN for the elderly is even more challenging because of an even greater requirement for usability. Such a PN must work for people who may not be accustomed to modern electronic devices or have lost their ability to deal with complexities. Further, it must also be usable for people who have reduced audiovisual capabilities and/or movement disorders, such as tremors in arms and hands.

This area of application poses a significant challenge since it requires ease of use for several very different groups of people, efficient and reliable communication, and also security. The system must be dependable, particularly in emergency situations. Privacy is another complex issue that cannot be neglected. While the elderly person wants a fast response in emergencies, he may not want to be monitored in detail all the time by unscrupulous relatives or neighbors.

1.3.4   More Use-Case Scenarios

Obviously, we can imagine many more PN use-case scenarios and applications. Here is
a short list of some additional use-case scenarios.

Walking through smart buildings. While a person walks through smart buildings from
   room to room, her PN accompanies her. It interacts with building functions and controls
   lighting, enables access to restricted areas, and activates building devices. For instance,
   the PN can incorporate a large wall-mounted display where she can view an incoming
   video stream directed to her, which otherwise cannot be displayed properly on her PDA.
Business environment extended from the office to the car. A person leaves his office
   and gets into his car. A PAN is established incorporating a number of car information
   accessories (via the on-board car network) so that he can listen to his corporate e-mail
   text read by a computer, dictate, and send replies. This could be realized, for instance, by
   linking up and temporarily extending the person's PAN containing a 3G-enabled PDA
   with on-board speakers, microphones, and a voice-recognition and speech-synthesis
   system.
A tele-presence session. One or more video cameras and high quality displays that sur-
   round a person in the office and at home can be used to set up a video conference or
   tele-presence session with someone else. The devices are incorporated, automatically
   and invisibly, into the person's PN as he enters the office or sits down on a couch in his
   living room. They allow him to start up a tele-presence session via a PDA, for instance,
   in which he can have a virtual meeting with other people for business as well as for
   social occasions. Alternatively, a person on the move could carry around some high
   quality portable wireless screens and cameras. Again, this would involve the automatic
   establishment of a PN involving local and remote devices.
A remote babysitting application. Consider the case of a mother visiting a friend's house
   while her child is asleep at home. She might want to remotely watch and observe the
   child. She does this by using a PN consisting of some personal devices, for example a
   UMTS and Bluetooth capable PDA and a headset she carries with her, and a remote
   pair of eyes and ears in the child's bedroom at home. The latter consist of a digital
   video camera, a microphone, and a UMTS phone, forming a cluster of cooperating
   devices. But since the friend's living room is equipped with a wall display including
   speakers, hooked up to the friend's home network and accessible to authorized guests
   via a Bluetooth link into the home network, she might want to use these to observe the
   child instead of her PDA and headset.

A way to envisage how these scenarios could happen is as follows. An individual
owns a PAN, consisting of networked personal devices in his close vicinity, for example
attached to the body or carried in a briefcase. This PAN is able to determine its context
(e.g. where it is), interact and link up with devices in the environment or with remote
devices in order to temporarily create a PN. This PN provides the functionality (e.g.
office functions in the car) that the individual wants at that very moment and in that
particular context.

These scenarios highlight some of the potential application areas of PNs. More scenarios
that reflect the vision of PNs have also been defined elsewhere (Jacobsson et al. (2004);
MAGNET (2005g); Niemegeers and Heemstra de Groot (2003)).

1.4   Federations of Personal Networks

The services and resources of a PN need not be confined to a single user. There are many situations in which it may be desirable to extend the boundaries of a single PN. A PN federation (Niemegeers and Heemstra de Groot 2005) is an extension of the concept of the PN that allows resources to be shared among different PNs. A PN federation is defined as a temporal, ad hoc, opportunity- or purpose-driven, secure group-oriented network where the users may be the producers and consumers of the services, content, and resources. In principle, only a subset of the resources of each constituent PN is committed to the PN federation. Only those resources are visible to the members of the PN federation.

The cooperation of PNs gives opportunities for different types of group-oriented applications in health care, education, business, entertainment, emergencies and more. Examples are distributed classrooms, sharing resources amongst project members, cooperative inter-vehicle networks, emergency networks, gaming and family networks. We will discuss PN federations in detail in Chapter 10.

1.5   Early Personal Network Implementations

Since PNs were first proposed, work has been going on to develop an architecture and solutions for them. In this book, we will introduce this architecture and the solutions along with some alternatives. This work has not just been theoretical, but also practical. A large part of it has been devoted to implementing prototypes. At the time of writing this book, at least three PN prototypes have been developed by different research projects, in particular the European MAGNET and MAGNET Beyond (http://magnet.aau.dk/) projects and the Dutch Freeband PNP2008 project (http://pnp2008.freeband.nl/).

The very first prototypes, which were developed within the PNP2008 project, were designed to demonstrate and test potential PN applications. Only very limited support systems were developed. Some of the prototypes were used in trials with real users. Based on user feedback, we were able to better understand what was really needed and the PN concept evolved accordingly. One example is the Medicum prototype, which was a demonstration of PNs in a professional setting – the medical profession. It showed how PNs and PN federations could be used to easily and reliably tie devices together in an area where errors are unacceptable.

Later implementations, such as those developed by MAGNET as well as PNP2008, were far more complete. They contain a good amount of support for networking, security, auto-configuration, context awareness, etc. Here, the focus was on the PN support systems and on testing them.

All these implementations clearly demonstrate that PNs can become a reality and that it need not take long. For all the details of these prototypes, we refer the reader to Chapter 11.

1.6   Expected Impact

The true impact of a new concept, such as the personal network, and the technology that underpins it is difficult to gauge quantitatively. It depends on many factors, such as user acceptance, market conditions, technology roadmaps, and regulatory frameworks, which are beyond the scope of this book. They will determine when and to what extent PNs might become a reality.

It should be remarked that PNs that have been prototyped in the different projects but they only address the basic functionality of PNs. We cannot yet talk about full-fledged PNs. The potential of PNs will grow as more advanced features, such as resource virtualization, context awareness, and cognition, become available for developing applications that go beyond what is presently possible. What we can point out, however, are the qualitative impacts that PNs might have in different domains.

PNs are based on personal devices, many of them consumer products from different manufacturers. Essential for the integration of these devices into a PN is that they can be software-enhanced to become PN-capable. One should be able to download and install PN software that incorporates the necessary PN protocols and functionalities into these devices. Therefore, it is necessary to define a core set of standards, an endeavor that has already begun.

A faster introduction and market penetration might take place in the professional domain. An example is the public safety and security domain, where professionals, such as firefighters, policemen, environmental specialists, ambulance personnel, security specialists, etc., might be equipped with specialized PNs to enhance their personal capabilities and allow them to federate with other professionals handling a particular incident. Other examples can be found in health care, for example (ETSI 2009). The community of players that have to agree on a common approach is much smaller and the urgency for adopting PN-like technology to increase professional capabilities and cooperation is much stronger. Moreover, the additional cost of making devices PN-capable may be small compared to the cost of professional equipment. Hence, an initial penetration of PN technology in these sectors could start in the short term, perhaps within a few years.

From the point of view of the consumer product manufacturers, PN technology should be seen as a product enhancement that allows devices to be embedded in a much more powerful distributed environment. This should extend the usage of these devices in space and time, and enable new distributed applications not yet foreseen. This in turn could be a sales argument for PN-capable consumer devices.

A significant impact will, we expect, be caused by the user experience created by PNs. The fact that a user has access, in a seamless way, to his personal devices and their services, wherever he is and with minimal user intervention, should be an unrivaled experience, especially given the growing numbers of ICT devices and services that are surrounding us.

PNs will be an enabler for the development of new distributed applications, exploiting the combined power and synergy of all the personal devices the user owns. In the light of the rapidly growing number and heterogeneity of these devices, from simple sensor and actuator devices to sophisticated computing systems, this may be a significant boost to the ICT sector in a new domain of applications and services.

PNs may also create new business roles for PN and service providers. This has been explored in the PNP2008 project (PNP2008 2008b,e), where the role of a PN provisioning party was defined, in conjunction with a supporting PN architecture. In such an approach, the provisioning of a number of the PN functionalities is outsourced to a trusted PN and/or service provider.

The PN concept and some of its supporting technologies can also be used in different domains. One can think of the concept of personalization and cooperation, and the supporting functionalities proposed in PNs as organizational principles for any future complex network that supports the functioning of one particular entity. Examples are networks for managing smart homes or buildings, for managing large vehicles such as ships, and for managing industrial structures such as power plants. To a certain extent, the principles of PNs could alleviate some of the issues raised by the Internet of things (ITU 2005).

Some of the technologies developed for PNs might be adopted in other ICT systems. Examples include the concept of personalization, which may be used to build grid-like systems in which secure cooperation among different entities is needed, or the self-organizational principles of PNs to form various types of overlay networks (Hoebeke 2007).

Finally, it is not only expected that PNs will have an impact on standardization, but also necessary. This process has started and may lead to a profile of standards for building PNs so that equipment from different manufacturers can fully coexist in a PN. This is likely to consist of a mix of existing standards and newly developed ones that together prescribe what is needed to implement PNs. This process is currently ongoing in an editing group within Ecma International (http://www.ecma-international.org/memento/TC32-PNF-M.htm).

1.7 Summary

In this chapter, we have introduced the vision of personal networks and how it creates the opportunity to exploit the possible synergy of the many personal devices a person will own in the near future. We discussed how the computer industry and the telecommunications industry have converged and now need to work together more closely in order to enable new applications that will be easy to use. Internet technologies and wireless communication will together allow seamless communication between a person's devices. Better computational capabilities allow for more intelligent and exciting applications. However, in order to achieve this, all devices need to be extended with software that allow them to seamlessly cooperate with each other. Personal networks were introduced to achieve this.

We highlighted the possibilities of PNs with some use-case scenarios. The scenarios show that PNs are beneficial in a large range of different situations, such as home networking, business, and health monitoring.

We also introduced the concept of federations of PNs. This broadens the user-centric concept of PNs into a group-centric concept. Federations allow the sharing of resources among different users and their PNs. The cooperation of PNs gives opportunities for different types of group-oriented applications in different areas, such as health care, education, business, entertainment, and emergency response.

Finally, we discussed current PN implementations as well as the potential impact of PNs. We discussed how a concept such as personal networks affects the development of new technologies and how the concept can become a reality, and argued that standardization is vital for the success of personal networks.

2

Personal Networks User Requirements

Before developing new solutions for personal networks, it is important to understand what exactly needs to be solved and what has already been solved. It should be clear that in order to realize the concept of personal networks, new solutions are required that can accommodate personal services and applications over a dynamic communication environment. To better understand what needs to be achieved, we discuss a set of important user requirements that need to be met. The requirements listed here are evolved versions of Niemegeers and Heemstra de Groot (2003) and PNP2008 (2006). Together, these requirements capture the total vision of personal networks, which also means that the requirements sometimes go beyond what this book will discuss.

These requirements are described at a very high level, hence it is impossible to precisely define if a requirement has been fulfilled. It is nevertheless important to try to formulate requirements to the degree that they can direct the research and standardization toward the relevant issues.

Sections 2.1–2.10 contain the user requirements that we consider important. They have been grouped into eight categories, with each category containing a number of related requirements. However, we do not aim to identify every single individual requirement as this is neither possible nor important at the moment. Section 2.11 highlights the link between the requirements and our story about Jane. Finally, Section 2.12 summarizes this chapter.

2.1 Ubiquitous Networking

Since PNs are mainly about communication, this requirement should not be a surprise. The devices surrounding a user should form a private personal area network (P-PAN) that enables communication using available wireless communication technologies. Both current and future WPAN technologies should be supported. Furthermore, the connectivity of the P-PAN must be extendable to devices beyond the close vicinity of the user by means of infrastructure-based wireless access networks, such as UMTS, WLAN hotspots, and WiMAX. PNs must be able to use any type of access technology and therefore be as

Personal Networks: Wireless Networking for Personal Devices   Martin Jacobsson, Ignas Niemegeers and Sonia Heemstra de Groot
© 2010 John Wiley & Sons, Ltd

independent of the infrastructure as possible. Regardless of network or device type, communication must be possible between any device belonging to the user whenever there is connectivity at the link layer. Hence, PNs should support a heterogeneous network environment by integrating all present network types into one ubiquitous network for the user.

From the scenarios in Section 1.3 we learned that communication with devices belonging to other persons and non-personal devices is also crucial. Ubiquitous communication over heterogeneous network environments with others, regardless of the geographical location of the devices, must therefore be possible as well.

Since many of the devices will be wearable or otherwise mobile, it is absolutely necessary to be able to deal with mobility. As devices roam through different networks, their communication links may break or new links become available. A PN must be aware of these events and adapt accordingly so that ongoing communication can be sustained.

All these networking issues need to be supported in a ubiquitous way, meaning that only minimal user intervention is required. All networking mechanisms must happen without the knowledge of the user. The PN needs to be able to establish and maintain itself on its own. In other words, PNs must be self-organized.

2.2 Heterogeneous Hardware Constraints

PNs will consist of a wide range of different mobile and stationary devices, wireless technologies and networks. They must operate efficiently in such heterogeneous environments. They should, for instance, be able to switch communication paths between different devices, links, and networks to achieve the best possible performance even when the number of devices and the amount of traffic is becoming large. We must make sure that devices that rely on battery power do not have to carry a load that limits their autonomy, and that utilized devices have sufficient computational power, memory, bandwidth, and other capabilities.

It is true that future technologies will bring us yet more computational power with less energy consumption and smaller devices with a more robust design at a cheaper price. Further, battery technology is also improving and new alternative energy sources for mobile devices are becoming available. The smallest devices that are thought to be needed as full participants in a PN are not the simplest. Sensor devices or similar have such tight hardware constraints that specially developed techniques are required. On the other hand, it is conceivable that such simple devices do not themselves need to become full members of a PN. Instead, they can be incorporated via more powerful devices that are full-fledged members of the PN. An example might be a sensor network, where the sink device, which is more powerful, is a PN member, but the sensors not.

Therefore, we can require more capabilities of devices that need to fully participate in a PN. However, the PN mechanisms must still run on battery-powered devices and extend the life of the battery to match the required system autonomy. The simplest devices that we would consider for PNs are wireless headsets, wristwatches, and other wearable devices. Currently, all these devices run on batteries that need to be recharged or replaced after some time. Using today's technology, it is acceptable for a device in a PN to be able to run for one or a few days before needing to be recharged again. Hopefully, developments in low power electronics and improved battery technology will alleviate this problem in the future.

2.3   Quality of Service and Reliability

Several potential PN applications have high demands for end-to-end quality of service (QoS), such as interactive applications and voice and video conferencing. The entire system should meet the demands of these and other applications with respect to QoS. Thus, parameters such as bandwidth, bit error rate, and latency should be considered in the routing and mobility management of a PN. The PN must be able to select communication paths that meet these expectations. In some cases, different network technologies must cooperate and QoS demands in each of them need to be fulfilled to meet the end-to-end demands of the applications.

The dynamic behavior of mobile wireless systems calls for very efficient adaptability to meet the demands of the user. The mobility management must be fast enough to respond to events such as broken wireless links, changes in link quality, or malfunctioning devices. If this is not the case, then the QoS requirements will be violated, making PNs useless for many important applications.

Another important requirement for PNs is reliability. Health applications, such as the ones outlined in Section 1.3.3, crucially depend on the PN. Given the unreliability of mobile and wireless systems, this is a challenging requirement. PNs should therefore not depend on a single network, but exploit the availability of many different networks and technologies at the same time to reduce the risk of being completely without communication possibilities at any given time. The level of reliability required by the applications should dictate how proactive a PN should be in finding and keeping backup links as this may imply extra energy consumption and perhaps cost.

Reliability can also be about instant data access and prevention of data loss within a PN. Important data should be backed up automatically to enable access to it at any time as well as protecting it from being lost. Otherwise, this may cause permanent loss of important data when devices are lost or break.

2.4   Name, Service, and Content Management

A PN provides a network architecture to support applications and services. However, applications and services still require additional software support to make it easy for developers to build them. This will also indirectly lead to applications and services that can better meet all user requirements. However, technical aspects of the network mechanisms should be hidden from both the user and the applications. This makes the system more integrated and at the same time it becomes easier to build applications and services for PNs.

Techniques to hide irrelevant aspects of the network layer include naming solutions as well as service discovery and management. Naming is needed to hide addresses and other irrelevant details of the network layer. Names can have meanings for users and give a human-understandable handle to relevant objects, such as devices, services, resources, and other objects. Naming is therefore very crucial to make PNs user friendly. The names can be assigned by the user to give an extra personal touch and in order to better organize the resources within a PN. Furthermore, network addresses may change, but names will remain unchanged until the user changes them. It is therefore better to use names to identify the various objects that users will see in a PN.

To achieve as much self-configuration as possible, it is better to make use of service abstraction. A service is an entity that offers client applications something useful through a known interface. The type and the capabilities of the service are described in a standardized way (Richard III 2001). These descriptions can be used by a service discovery mechanism to enable the applications to easily find available services and select the most appropriate one. Furthermore, a good management framework is also required that can manage not only the services, but also the clients and their service sessions. The task of management is to control the service usage when the network situation changes so that the clients and the services can operate optimally.

Content files, such as documents, music clips, and video clips, should also be managed by the PN. Given the wide scope of the applications expected to run on PNs, one has to deal with a great variety of different types of content, with different characteristics in terms of importance, required storage, sensitivity, the time dependency of its value, etc. At the same time, the personal devices that store content can also have vastly different capacity and speed. Add to this the dynamics of a PN and it becomes obvious that a content discovery and management facility should be part of the PN. Such a facility must make sure that enough backups of the content are kept and that files are accessible when needed.

2.5   Context Awareness

Context information is anything that can characterize the situation of an object, such as a person, a device, or a network (Abowd et al. 1999). This information is valuable since it can influence the behavior of the PN applications or the PNs themselves. The more information available to an application, the better that application can respond to the user and the situation. While this additional information may, for many applications, not be absolutely crucial, it is still desirable in order to meet the user's high expectations.

We would like devices and applications to be intelligent, to properly predict user intentions, and to automatically adapt to a changing environment. It is therefore necessary to implement a context information framework that can discover, process, and distribute relevant context information. Furthermore, the PN and its applications must be able to make proper use of this information. That is, both the PN and its applications must be context-aware.

2.6   Being Cognitive

Since a PN will be a very dynamic entity, evolving with its user in terms of applications, services, and resources, we would like it to be able to adapt automatically to new circumstances without somebody having to reprogram it. This involves more than self-management. It requires a system that learns from its experiences and is able to evolve autonomously. PNs introduce into the design the dimension of choice of resources and objectives. The number of entities involved should be scalable to large numbers as more and more personal digital devices serving the user become available. The short-term dynamics due to mobility, radio link behavior and the changing state of devices and applications can be considerable. All this leads to very complex problems to be faced by the PN design – problems that may be intractable and for which algorithmic solutions are not feasible.

Therefore, we envisage cognitive solutions, whereby the PN is able to observe many parameters, regarding context, system state and decisions made, and build up the history of these. Then the PN can detect patterns and trends, predict, learn from decisions, and thus improve its decision making processes. In other words, it can exhibit cognitive behavior at various levels and functionalities in order to better serve the user.

This cognitive approach is, of course, in itself complex. However, we feel that in the PN context it will become feasible. The reason is that the large data collection, computing, and storage resources needed for the cognitive functions can easily be part of the PN. What is required is that, for instance, a home computing system is part of the PN. Given the expectation that Moore's law remains valid for some time to come, we can expect such PC-based systems to have the required power at consumer prices.

2.7 Security and Trust

The new characteristics and possibilities offered by systems like PNs lead to new security and trust problems that need to be properly addressed (Stajano 2002b). PNs can only succeed if people trust them, but unfortunately PNs have extra vulnerability because of their mobile and wireless nature. In the world of mobile communication, ICT security meets traditional security and this opens up a new world of problems in the security domain. The ad hoc nature of PNs means that a person's PN will encounter many unknown parties but must remain properly protected from those it does not trust.

The main challenge to security for PNs lies not in the security algorithms or security protocols. There is a rich variety of security solutions that PNs can leverage. The problem is rather to find a way to implement trust among persons, devices, and networks that is both powerful enough to contain all the necessary details and at the same time comprehensible for a normal user. Trust models and their security systems may become too complicated, with the consequence that the users are severely bothered by them or even fail to sufficiently understand them (Balfanz et al. 2004). At the same time, they might fail to protect the right things, because it is no longer clear what to protect and against whom in a world of mobile and ubiquitous communication (Stajano 2002b).

Even so, a security system is needed and it must protect the PNs and their users from unauthorized usage. It is important to note that the security system must be an integral part of the system design and not an add-on. The security must also work when devices are stolen or compromised. Making consumer products, such as PN devices, tamper-proof is too costly and difficult (Stajano 2002a). Hence, another solution must be found that can maintain the security of a PN when devices are lost or stolen.

2.8 Privacy

As more and more information in our lives is being digitized, privacy is becoming an even bigger issue (Garfinkel 2000). Privacy is about protecting the data kept inside a PN as well as preventing the possibility of tracking a user's activities through her PN. First of all, traffic between a user's devices must always be encrypted. However, this is not enough. In the world of mobile communication, there is an increasing risk of theft of devices. We already know that a stolen or forgotten laptop or PDA may lead to confidential information falling into the wrong hands. Unfortunately, with personal networks, it becomes even more important because of the extra capabilities a PN device will have. Hence, we need to build a system that minimizes the impact in the event of lost or stolen devices.

Another privacy problem is that being recognizable by the devices you carry can also be an intrusion into a person's privacy, since this information can be used to track a person's movements and activities. This can be done since many wireless devices expose their identities in the form of link layer addresses or other unique and fixed identities. Anonymity is therefore needed in PNs. In terms of wireless networks, it means that a device must never expose its identity or anything that can easily be linked to its identity to non-trusted parties (Schmidt 2002). This is important since we are likely to carry the same devices all the time and this can be used by unauthorized persons to track the movements and activities of users. However, preventing all types of identity exposure is an almost impossible task. What we can do is avoid the most obvious pitfalls, such as transmitting fixed link layer addresses or other fixed identities unencrypted. If that is done, more sophisticated methods will be required to track someone.

2.9 Usability

With PNs, each person may have several embedded and wearable computers in addition to present-day devices such as laptops and PCs. At the same time, PNs should be for everyone and not just for experts and technology freaks. In fact, anyone, including children, pensioners, and the physically impaired should be able to use a PN. Furthermore, a PN must support the user in his daily activities in an efficient and pleasant way. This means that usability is one of the most important requirement for PNs.

A user must be able to easily create a new PN and add his personal devices to it. When this is done, the PN must be self-configured and its devices must be able to communicate seamlessly with each other without requiring complicated user intervention. It must not be necessary for the user to configure any network settings, such as addresses or default routes. All networking solutions must be self-organized. Further, the PN devices must also automatically adapt to new situations without user involvement. The PN applications should themselves detect their own settings and operate over any type of network in order to carry out their tasks. That is, a PN and its applications must be able to operate without directions from the user. At the same time, this must be done in a secure way and with the user still in charge of all his devices. This is a delicate problem and serious care must be taken when designing these types of systems.

The best way to achieve usability is actually to design a simple and intuitive system that is still capable of fulfilling its other requirements. The idea is for this to be so easy that the user easily can understand how it works and create an accurate mental model of the system. Whenever the system does not work according to the user's wishes, he knows why and what needs to be done in order to make it operate correctly. In this way, the user will be in charge of the system and the system will never do something unexpected. On top of this simple architecture, smart and complex components can be designed for specific issues.

Especially from a security perspective, a simple and understandable design is crucial (Dourish et al. 2004; Whitten and Tygar 1999). Users who do not understand the system will probably not understand that they are vulnerable to malicious attackers and no user interface can perfectly hide such a flawed system design. In the worst case, the security system is so complex that the user becomes annoyed and ignores security warnings as well as trying to disable or circumvent it by, for instance, using simple-to-guess passwords or even no passwords. This will allow access to everything for everybody. It is therefore important that careful consideration is given to the issues related to usability and security. Otherwise there is a high risk of the user becoming the 'weakest link'.

2.10 Other Requirements

These user requirements are by no means exhaustive. There are many more requirements depending on the target group, applications, and environment. In addition to user requirements, business-related requirements that various commercial personal network stakeholders may have are also not included, nor are requirements related to legislation and regulations. An attempt to cover a more complete set of requirements using user scenarios, use cases, and other approaches has been made by IST MAGNET (MAGNET 2005b).

Furthermore, the requirements that we have identified in this chapter as well as the requirements identified elsewhere should be subject to an iterative and continuous process of improvement. This should eventually lead to several sets of proper requirements that can be validated. As PNs continue to evolve, it is also anticipated that more requirements will arise.

2.11 Jane Revisited

The most important features offered by PNs are the enhancement of the mobile experience and the possibility of new services and applications. All this should be possible in the near future and without waiting for additional breakthroughs in technology. That these requirements are relevant to the PN concept can also be illustrated with reference to Jane, who was introduced in Section 1.3. Again, this is not an attempt at a rigid and fully exhaustive requirement analysis, but merely a motivation of the PN concept.

> *That Jane needs ubiquitous connectivity (Section 2.1) is obvious. To access files and use applications in other locations, Jane needs network connectivity. For streamed media and health-related services, she needs quality of service and reliability (Section 2.3). Furthermore, she needs this to be hassle-free, that is, she requires auto-configuration and self-organization (Section 2.9). Since Jane may not want to reveal certain files within her PN or within her company network to her clients, etc., she needs authentication and security (Section 2.7) as well as privacy (Section 2.8). For efficient handling of her files and other content, she needs content management (Section 2.4). Context awareness (Section 2.5) and cognitive approaches (Section 2.6) will improve the system's response to Jane and her situation.*

2.12 Summary

In this chapter, we have formulated a small set of high level user requirements that need to be fulfilled in addition to the current state of the art. The requirements include such issues as ubiquitous networking, dealing with heterogeneous hardware constraints, QoS, reliability, naming, service management, context awareness, security, trust, privacy, and usability. We argued that solutions that do not meet these requirements will not be able to meet the user expectations of personal networks. However, we envision that further requirements will arise as PNs evolve and the telecommunications and computing landscape change.

3

Trends in Personal Networks

Many technologies have been proposed in the area of personal communication, but there have been very few attempts to create a complete and integrated solution that caters for all present and future personal communication needs. In this chapter, we present earlier and current work aimed at either analyzing future requirements or proposing and/or building integrated solutions for personal communication. In assessing the related work we will refer to the user requirements discussed in Chapter 2. We only discuss the more complete attempts here and not solutions that only address a particular aspect. Of the latter, there are quite a number. It is clear that personal networks will build on many of those existing solutions and technologies. We will introduce the relevant ones in their appropriate context in the later chapters.

Since the content of this book is, to a large extent, based on research that we, together with others, conducted in IST MAGNET, IST MAGNET Beyond (http://magnet.aau.dk/), Freeband PNP2008 (http://pnp2008.freeband.nl/), and IOP GenCom QoS for PN@Home (http://qos4pn.irctr.tudelft.nl/), the results of those projects are not discussed here, but are reflected throughout the book.

In this chapter, we start by taking a brief look at the wireless communication trends and technologies that are likely to have the most impact on the development of PNs in Section 3.1. We then look at the developments in ad hoc networking in Section 3.2, the vision of the Wireless World Research Forum in Section 3.3, and ubiquitous and pervasive computing and communication in Section 3.4. Next, we look at the relevant results of specific major projects. We look at the IST projects Ambient Networks in Section 3.5, PACWOMAN and SHAMAN in Section 3.6, Mobile VCE personal distributed environment in Section 3.7, MyNet in Section 3.8, and the P2P Universal Computing Consortium (PUCC) in Section 3.9. Finally, we consider some trends that are likely to have a considerable impact on PNs in Section 3.10, before concluding the chapter with a summary.

3.1 Wireless Communications

The state of the art in wireless communications is covered in many books. A particularly up-to-date view of the ongoing research and future evolutions is provided by the research-oriented WWRF series (David 2008; Tafazolli 2004, 2006) that will be discussed in

Personal Networks: Wireless Networking for Personal Devices   Martin Jacobsson, Ignas Niemegeers and Sonia Heemstra de Groot
© 2010 John Wiley & Sons, Ltd

Section 3.3. For a more introductory view of the state of the art, a good reference is (Garg 2007).

There are, in the realm of wireless communication, two major developments that are strong enablers for PNs. The first is the ubiquitous availability of wireless infrastructure networks, possibly augmented by ad hoc or mesh networks to extend their reach, and the second is the availability of inexpensive short range wireless technologies that allow the interconnection of virtually any device. Together these developments create a communication web that will allow a cost effective and seamless interconnection of all devices locally but also globally via the Internet, a prerequisite for forming PNs. These technologies address a wide range of communication requirements, in terms of range, data rates, and quality. Let us briefly survey some of the more important wireless trends that will play a major role in the development of PNs.

Wireless infrastructure. After the enormous global expansion of cellular communication based on GSM and the various 3G technologies, we can expect a rapid upgrading of these networks to keep up with the bandwidth and quality demands imposed by the growing Web applications and multimedia communications. Worldwide, this effort is borne by Long Term Evolution (LTE) (Sesia et al. 2009) and is the subject of standardization by the 3rd Generation Partnership Project (3GPP) (http://www.3gpp.org/), an alliance which is creating the 4th Generation (4G) standard LTE Advanced (Rumney 2009). There are also technologies that are competing with the present 3G systems such as UMTS, in particular WiMAX, based on IEEE 802.16 and the ETSI HiperMAN standards (http://www.wimaxforum.org), which provides similar services in regions of the world where UMTS is not penetrating. However, these technologies will ultimately be incorporated in 4G systems. Another complementary development is the WLAN-based networks that cover complete urban areas (Brik et al. 2008). At the same time, the emergence of femtocells (Zhang and de la Roche 2010) will allow the reach of cellular infrastructures to extend into buildings and offload the main network in dense urban areas. Complementary to the cellular infrastructure is also mesh networking where wireless routers form ad hoc mesh networks (Methley 2009) relaying each other's data streams and messages to and from the infrastructure. The net result is that, at least in developed economies, Internet connectivity is becoming available to mobile devices in any geographical location. The 4G standardization activities are working on integrating all these infrastructural technologies into a seamless wireless infrastructure for global communication.

Short range. A second enabler for PNs is inexpensive short-range communication technology. Many devices that will be part of a PN will be small, inexpensive, and often wearable. For reasons of cost and energy consumption, it might not be appropriate to equip them with cellular or longer range radios. Short-range radios are more appropriate to connect such devices to each other and to more capable devices that connect to the Internet via more powerful (long range) radios. Short range radios are also ideal for connecting personal devices that are close enough to each other.

There has been a proliferation of short-range radio technologies in the ISM band, one of the best known being Bluetooth. The IEEE 802.15 working group for wireless PAN (WPAN) (http://www.ieee802.org/15/) is the most prominent standardization body that defines such technologies. The standards aim at ranges of around a few to

10 meters, suitable for in-house and in-car communication and for body area networks (BANs). A range of standards is defined corresponding to different data rates, quality, and energy requirements for different classes of applications. Examples are IEEE 802.15.1 used by the Bluetooth alliance (http://www.bluetooth.com) in many personal devices, the low power, low data rate IEEE 802.15.4 used by the Zigbee alliance (http://www.zigbee.org/) for sensor systems and RFID, and the high data rate (20 Mbps or greater) IEEE 802.15.3 used by the WiMedia alliance (http://www.wimedia.org/) aimed at multimedia systems. These technologies and the corresponding standardization are still evolving to match the evolution of user requirements, such as the transmission of HDTV streams (order of 10 Gbps) and the imminent USB 3.0 (5 Gbps) standard. New unlicensed bands, in particular the 60 GHz band that offers a lot of bandwidth, are targeted and various ultra wide band (UWB) techniques are proposed. An important goal is to be able to replace the increasing number of cables used in computing and entertainment systems by more convenient wireless links, such as wireless USB and wireless HDMI.

An issue of growing concern with the proliferation of wireless devices, in particular operating in the ISM bands, is their mutual interference. There are two types of interference: *intra*-technology interference, where many networks using the same radio technology, such as IEEE 802.11, interfere; and *inter*-technology interference, where networks operating according to different standards disturb each other. One example of the latter is Bluetooth and IEEE 802.11. Avoiding inter-technology interference and making multiple radio networks that operate according to different standards in the same frequency bands coexist is an issue that has not been fully addressed. Examples of recent work in these areas include Pollin et al. (2006) and Yuan et al. (2010). Furthermore, the IEEE has also dedicated its 802.15.2 coexistence standardization activity to this topic.

Both short range technologies and wireless infrastructure have their merits and drawbacks. PNs will need to use both and be able to exploit their capabilities to the best possible extent. An important task of PNs is to integrate them and make them work together seamlessly. At the same time, PNs must be able to work with the many wireless technologies, both current and future, in order to guarantee their usefulness.

3.2   Ad Hoc Networking

Ad hoc networking is not really a new topic. Its main foundation started decades ago under the term Packet Radio Network (PRNET) and was developed for military purposes (Kahn et al. 1978). Around 1996, the academic community began to show greater interest in the topic with non-military applications becoming more important, such as emergency relief networks, home networking, community networks, and WPAN. From there, the research interest saw remarkable growth for some years, which made ad hoc networking a serious research area within wireless and mobile communication.

Ad hoc networking is all about quickly and automatically setting up an unplanned wireless network. Wireless devices should be able to automatically find each other and establish a network. When devices and their networks become mobile, this is even more

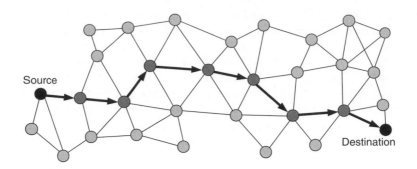

Figure 3.1   A multi-hop MANET with a path.

important. Hence the birth of the frequently used term mobile ad hoc networks (MANETs) (http://www.ietf.org/html.charters/manet-charter.html).

A MANET consists of mobile devices with wireless communication capabilities that can move around freely and still communicate with each other without infrastructure support. Due to limitations in the utilized wireless communication technology, it may not be possible for two MANET devices to communicate directly. Instead, devices in-between will assist by relaying the packets so that they can reach their final destinations. This is called multi-hop communication since a single packet must be retransmitted in several hops to reach its destination. Figure 3.1 shows a MANET that has a multi-hop path between a source and a destination.

When an ad hoc network is not fully connected, we need multi-hop communication. In order to find the hops that can connect the source and destination, routing is required. Many routing protocols have been designed specifically for multi-hop MANETs (Broch et al. 1998; Perkins 2001). In fact, routing has been the most covered aspect of ad hoc net-working and is still an active area of research (Chakeres and Perkins 2009; Clausen et al. 2009b). Examples of other frequently studied areas in ad hoc networking are network-wide broadcasting (Stojmenović and Wu 2004; Williams and Camp 2002) and security (Yang et al. 2004).

Ad hoc networking is an ideal technology for PNs as it offers self-configured, self-maintained, and self-organized networking. As a consequence, ad hoc networking will be an important building block for PNs.

3.3   WWRF Book of Visions

The Wireless World Research Forum is a global forum of academia, research institutes, and industry in the area of wireless communication. It includes manufacturers, network operators, service providers, and other related companies. The forum was founded in 2001 and provides a global platform for discussion of results and exchange of views to initiate cooperation toward systems beyond 3G. One outcome of this joint forum was the Book of Visions, first launched in 2000 (Wireless Strategic Initiative 2000) under the Wireless Strategic Initiative (WSI), a research project sponsored by the European Union. The Book of Visions later appeared in book form under the title *Technologies for the Wireless Future* (David 2008; Tafazolli 2004, 2006).

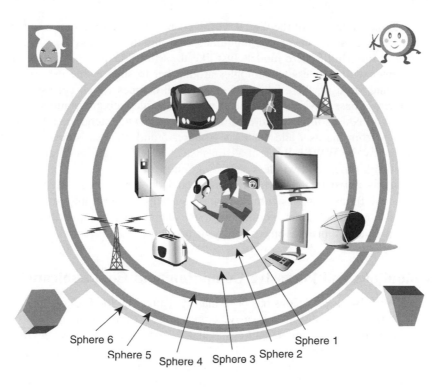

Sphere 6
Sphere 5   Sphere 4   Sphere 3   Sphere 2   Sphere 1

Figure 3.2   The MultiSphere model of the WWRF Book of Visions (Tafazolli 2004).

The idea behind the Book of Visions was to bring together the experts in this field to gather ideas and outline visions and challenges for the research and development of future wireless communication systems. Among many ideas, the so-called MultiSphere model was proposed to support further definitions and work on complex mobile communication concepts and ideas. Based on the evaluation of some usage scenarios, this model was defined to reflect the key characteristics of future communication scenarios. A graphical representation of the model is shown in Figure 3.2. The model is used to better understand the future of wireless communication and identify areas that need research.

The MultiSphere model acknowledges the importance of usability by placing the user at the center surrounded by various wireless communication systems that work together on the user's behalf. These systems are divided into spheres. The PAN that connects a person's handheld and wearable devices is the innermost sphere, sphere 1. Sphere 2 consists of elements in the immediate environment with which the PAN devices can communicate. Sphere 3 consists of other nearby persons and more complex networks, and sphere 4 consists of the mobile networks (e.g. GSM and UMTS) as we know them today as well as future wireless wide area mobile networks. At this point, it is important to have interconnectivity among all the wireless technologies in the inner spheres, and this is placed in sphere 5. Efficient and seamless integration is seen as very important since it must be possible for all wireless devices and persons to communicate with any other wireless device. The outermost sphere, sphere 6, is the so-called CyberWorld. Here, the wireless world meets the rest of the digital world, such as the Internet. In the CyberWorld, we may interact with smart agents, communities, and digital services.

PNs partly fit the MultiSphere model. Just like the MultiSphere model, they place the user at the center with wireless technologies around her to support the user in her daily activities. Around the user, there is a PAN of personal devices which corresponds to the innermost sphere. These devices can interact with the immediate environment (spheres 2 and 3). Further, a PN uses infrastructure networks (sphere 4) to extend the PAN to devices physically away from the user and her PAN (spheres 5 and 6). PNs address many of the issues identified by the WWRF Book of Visions, such as self-organization and integration of various network types, and wireless service architectures.

It must also be noted that the MultiSphere model does not cover all the aspects of either PNs or any future wireless communication systems, for example, it does not give much insight into human to human communication using wireless communication systems. Neither does it give any insights into security or privacy.

3.4   Ubiquitous and Pervasive Computing and Communication

While the WWRF Book of Visions has been mainly defined by the European telecommunications industry, ubiquitous and pervasive computing is a more American drive aimed at future computing and communication and has its origins, to a large degree, in the computer industry. The term was coined by Mark Weiser (Weiser 1991) of Xerox PARC in 1991. He noted that 'people find a walk among trees relaxing and computers frustrating' and suggested the aim of making computing more ubiquitous, that is, intertwining computers in everyday life and making them 'vanish' into the background. Computing should be embedded into everyday objects to interact with people and to enable them to move around. This would lead to enhanced user interactions since computers would disappear from our focus and become non-intrusive. Devices should be tailored to suit particular tasks. Further, they should be enabled to sense changes in the environment and automatically adapt and act based on these changes as well as user needs and preferences. To be really usable, these ubiquitous computers also need to be interconnected so that they can communicate with each other and thereby cooperate to meet the needs of users.

Ubiquitous computing is very different from personal networks in one major aspect. In the view of ubiquitous computing, computing devices are seen as commodity items that serve any user. They are meant to be shared by everybody. In fact, a personal laptop is even seen as a failure. No one should need to bring a computer as there should be ubiquitous computing objects available everywhere to be used by anyone. However, the current trend is toward increasing numbers of personal devices. The most successful of these is without doubt the mobile phone, followed by laptops, personal digital assistants (PDAs), navigation systems, e-book readers, portable storage devices, and many more. Today, these devices are not only carried for their functionality, but also as personal attributes and status symbols. We love to have them and we love to give them a personal touch by giving them personal background images and ringtones. The concept of PNs acknowledges that some devices are personal and mainly used by one person. At the same time, there are ubiquitous computing devices in the surroundings that these personal devices can interact with.

Much of the research that has taken place to bring about the vision of ubiquitous computing and communication is also very useful for PNs. This includes, but is not limited to, ad hoc networking, sensor systems and networks, and ubiquitous user interaction.

Besides ubiquitous computing and communication, the terms pervasive computing and communication are also frequently used. Another, almost identical, vision is ambient intelligence (Aarts and Encarnação 2006). This was first defined by the European Commission's Information Society Technologies Advisory Group (ISTAG) in 2001 (ISTAG 2001) as a way to stimulate European research in this area. The Ambient Networks research project, discussed in the next section, is an example of such research.

3.5   Ambient Networks

Ambient Networks (http://www.ambient-networks.org/) was a European research project that stemmed from the Wireless World Initiative (WWI), a spin-off from the WWRF. It was sponsored by the European Commission under the Information Society Technology (IST) 6th Framework Programme (similar to IST MAGNET and IST MAGNET Beyond). Its main objective was to create network solutions for mobile and wireless systems beyond 3G.

Most of the work carried out in this project concerns user devices and networks and their connections to access networks. Today there are many different types of networks available to users, and new network types are constantly being deployed. The main idea behind this project was to make each of these networks into ambient networks (ANs). The project offered a fundamentally new vision based on the dynamic composition of these ANs to avoid the growing patchwork of extensions to existing networks. This should provide the user with access to any network, through instant on-demand establishment of inter-network agreements.

A comprehensive prototype was developed within the project that gives users or networks the choice of using the appropriate radio technology automatically. They can switch between different flavors of 3G systems, WLAN, Bluetooth, or forthcoming 4G systems depending on what is the best network for a particular service or multimedia content. The project developed solutions for providing QoS sensitive multimedia services.

Ambient Networks is more about the linkage between users' networks and infrastructure networks and between the different infrastructure networks than about the users' networks themselves. However, these links are still important for PN communication and will help in reaching ubiquitous networking with QoS support and reliability. Ambient Networks solutions can provide seamless infrastructure support for PNs.

3.6   IST PACWOMAN and SHAMAN

Power Aware Communications for Wireless Optimised Personal Area Network (PAC-WOMAN) (http://www.imec.be/pacwoman/Welcome.shtml) and Security for Heterogeneous Access in Mobile Applications and Networks (SHAMAN) are two other IST projects that started slightly ahead of IST MAGNET. PACWOMAN worked mainly on WPANs and ad hoc networking. The networking environment was divided into three distinctive spaces (Louagie et al. 2003; PACWOMAN 2002). The first space was the personal area network (PAN), where personal devices can communicate with each other. The second space was the community area network (CAN), which consists of nearby PANs belonging to different people that wish to interact with each other. The last space was the wide area network (WAN), which provides each of the PANs with connectivity to remote devices. PACWOMAN's main area of research was link layer and medium access control for the PAN space. They identified the need to separate low and high data rate communication. Low data rate technologies are needed for sensors and other small devices with limited power. Such devices require simple and power-aware networking, something that is highly useful for any type of personal networking.

IST SHAMAN (SHAMAN 2002) focused on providing a security architecture for PANs. The basis for their architecture was a trust model (Gehrmann et al. 2002a) that describes the basic security relations between different PAN devices (components in the SHAMAN terminology). Each device is owned by one user who determines, by means of security policies, who can use it. The security framework covers both local communication within a PAN and global access to the infrastructure. With respect to a particular device, other devices are classified either as first party devices, trusted second party devices, or non-trusted devices. A first party device has the same owner as the device itself and therefore has the highest trust. This model is implemented with a personal certification authority (CA) (Gehrmann et al. 2002b) that runs on one of the owner's devices and hands out certificates in the form of public–private key pairs to all the user's PAN devices.

PACWOMAN and SHAMAN are highly relevant to PNs, and developed many relevant concepts: PACWOMAN addressed part of the ubiquitous networking, device hetero-geneity, and QoS requirements, while SHAMAN addressed usable security and trust requirements. However, one needs to combine the two into a single solution.

3.7   Personal Distributed Environment

This research project is part of the United Kingdom's Mobile Virtual Centre of Excellence (http://www.mobilevce.com/) and has defined a concept referred to as the personal distributed environment (PDE) (Dunlop 2004; Dunlop et al. 2003), based on a very similar vision to PNs. The PDE is an attempt to define a concrete architecture and implement solutions that are driven by this vision. It is important to point out that the PDE project and the MAGNET project originated independently and have been going on in parallel. The proposed architecture and solutions have similarities but are also very different in other aspects.

A PDE consists of a user's local and remote devices and services (Schwiderski-Grosche et al. 2005), as shown in Figure 3.3. At the center of the PDE is the so-called PDE server, which has a device management entity (DME). The PDE server is basically a normal server, somewhere in the Internet, and the DME is a database of all devices belonging to a particular user. Each device in a PDE stays in contact with the PDE server to update its location, capabilities, and services in the DME. This connection can either be direct using a secure tunnel or by means of a proxy service on another device in the same PDE. In this way, it is possible for a device in the PDE to use services on any other device in the PDE through the PDE server.

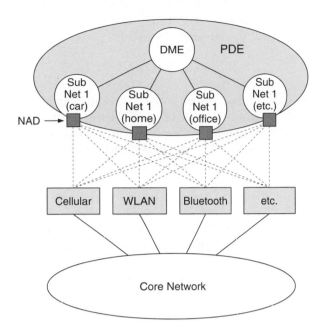

Figure 3.3 A personal distributed environment (PDE) with four subnetworks (Dunlop 2004).

While PDE focuses on service delivery over heterogeneous networks, it also covers many network aspects. Devices within the same PDE that can communicate with each other using a local communication technology form a so-called 'subnetwork'. Examples of subnetworks are PANs, BANs, home networks, and office LANs. Subnetworks are connected to each other through a core network using various access technologies and through so-called network access devices.

The PDE assumes that each subnetwork already implements the necessary network and security solutions. These local mechanisms may differ between different subnetworks and networking environments, but are kept unchanged. Instead, to make sure the PDE and its devices do not perform unauthorized tasks, a trust management system based on a trust engine that bridges the various trust and security systems in the various subnetworks is proposed. This system is part of the DME and is also responsible for trust and security with respect to devices outside the PDE.

The PDE is thus not a single homogeneous system. The benefit of this approach is to keep legacy solutions as they are and only provide integration when needed. However, it also makes the subnetworks quite different from each other. As a consequence, the user may be burdened with several different technologies, configuration possibilities, user interfaces, and the like. Further, the subnetworks will most likely not be compatible with each other and make it difficult for devices to move from one subnetwork to another. Nevertheless, the PDE concept addresses many of the requirements of personal networking. Notable exceptions are context awareness and privacy. From a usability and

networking perspective, we believe it is better to provide one common security and
network solution for all devices and networks in a PN.

3.8 MyNet

The MyNet project (Arvind and Hicks 2006) is a collaboration between Nokia and
Massachusetts Institute of Technology (MIT) and aims to study and develop a network
architecture, tools and applications for simple, secure, personal overlay networks. The
User Information Architecture (UIA) (Kaashoek and Morris 2006) and the Unmanaged
Internet Protocol (UIP) (Ford 2003) are projects within MIT. MyNet is based on these
projects.

UIP combines the self-management of ad hoc networks with the scalability of IP by
creating a self-organized overlay network for personal devices. UIA, on the other hand,
is intended to allow global interaction and sharing among information devices between
persons. The UIA protocols are the foundation upon which the rest of the MyNet project
work is built. UIA is based on two principles: (a) security is decoupled from physical connectivity; and (b) establishment of trust is based on social connectivity. This is achieved
by creating personal and private name spaces. Simple-to-use mechanisms that leverage
social relationships will allow a user to share access to their devices and resources using
these name spaces.

These projects stem from the peer-to-peer research community, but are highly relevant
as they focus on many areas that are important for PNs. These projects do not fully
support ubiquitous networking, but do address ease of use, self-organization, security,
and trust as well as naming management.

3.9 P2P Universal Computing Consortium

The P2P Universal Computing Consortium (PUCC) (http://www.pucc.jp/) is a collaboration between a small number of Japanese universities and companies, such as Toshiba,
NTT DoCoMo, and NEC. Its aim is to realize a seamless peer-to-peer (P2P) communications technology platform that enables the creation of ubiquitous services between
networked devices. The initiative has been running since December 2004, but until
recently very little has been published.

The goal of PUCC is very similar to that of PNs. With P2P overlays, they provide
seamless communication between IP networks and non-IP networks, such as home networks and sensor networks. A service platform provides seamless integration of services
and other higher layer functionalities. However, the network layer is kept as is without
any extra support, such as auto-configuration and self-organization. Figure 3.4 shows the
proposed PUCC protocol stack.

Currently, protocol specifications have been proposed for most of the core functionalities as well as for key applications. Developer kits for networked devices are being
prepared. A demonstration was given at the Consumer Electronics Show in Las Vegas
in January 2008. Despite the limited information published about PUCC, we believe it
addresses many personal networking requirements, including service discovery and management as well as security. However, similar to our view of MyNet, we do not feel that
the requirement of ubiquitous networking will be fully addressed by PUCC.

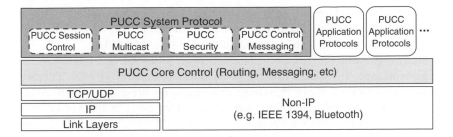

Figure 3.4   PUCC platform protocol stack.

3.10   More Trends

There are numerous other projects that touch on aspects of PNs. One early attempt is Universal Personal Networking (UPN) (Braun et al. 1993), a Siemens project in the early 1990s. At that time, WPANs were virtually non-existent and WLANs were very new. However, the aim of UPN was similar to the concept of PNs, but the existing technologies at that time were a limitation. Hence, UPN focused on infrastructure support for personal networking, device technology, and user interfaces. Taking place prior to the big breakthrough of the Internet, they put a lot of effort into Internet-like techniques but neglected security aspects. A more recent initiative from Siemens is their Life-Works (http://www.enterprise-communications.siemens.com/Open%20Communications/Our%20Vision-LifeWorks.aspx), which is a visionary concept of an unified communications experience for both business and private users. Under the umbrella of LifeWorks, Siemens develops products that aim at seamless convergence between fixed and mobile networks, new better services for mobile users, and ease of use. Now the entire telecommunications industry is showing more and more interest in this area. However, the current focus is more on businesses and business services.

IBM defined and showcased a concept called the personal mobile hub (PMH), which acts as a hub between a PAN and the infrastructure network (Husemann et al. 2004). It can connect and control the PAN consisting of a person's wireless devices and also interconnect them to servers in the infrastructure. To demonstrate the concept, IBM developed a health-related application that monitored heartbeat and blood pressure and gave an alert when certain thresholds were exceeded. Further, it could monitor whether a person took his medication and otherwise alert the person and/or caretakers.

In the academic world, it is also worth mentioning the work on personal networking by Robin Kravets's group at the University of Illinois at Urbana-Champaign. Among the solutions they worked on, there is one called the mobile grouped device (MOPED) (Kravets et al. 2001). MOPED is a system that represents a person's set of personal devices as one entity toward the Internet using only one single Internet address. That address is given to a proxy node that is always available through the Internet. It is the task of the proxy to keep track of all the other personal devices and how they are connected to the Internet and to each other. Personal devices that can connect directly with each other form what they call components. The components may then connect to the Internet and the proxy. Hence, MOPED provides a technical solution to achieve personal networking and its focus is clearly on addressing, routing, load balancing, and mobility. While it solves many important aspects, there are many more still left open, such as security, support for higher layers, and direct wireless communication between MOPEDs.

There are many more smaller projects or specific solutions that target selected areas of PNs. For instance, HP's CoolTown (Debaty and Caswell 2001) gives people, places, and things a presence on the web. These dynamic web presences can then be related to each other to form new interesting applications that connect the virtual world with the real world. Stanford's Mobile People Architecture (Maniatis et al. 1999) proposed an application-level mobility solution for mobile persons, since it is the persons that are the end-points and not the devices. It introduces a personal proxy that tracks the person and handles personal-level mobility aspects, including accepting incoming communication on the person's behalf, directing it to the correct device, converting the communication stream if necessary, and protecting the person's privacy. Both CoolTown and Mobile People are completely infrastructure-based and do not consider local communication and many other aspects of personal communication. However, they have brought useful concepts and partial solutions for realizing PNs.

Finally, it is also worth mentioning that 3GPP, in their drive toward all-IP networks (AIPN), recently started to consider use-cases similar to personal networks (3GPP 2009b). In fact, they use the term 'personal networks' for those use-cases which involve a person with devices in different locations that are interconnected using 3GPP-networks as well as non-3GPP networks. In 2008 they defined a personal network management service (3GPP 2009a).

3.11   Personal Networks and Current Trends

How close does the work presented in this chapter come in terms of providing solutions to meet the user requirements outlined in Chapter 2? The work on ad hoc networking, WWRF, and ubiquitous computing and communication we discussed was more targeted toward steering research and development rather than producing concrete solutions. While research done within any of these areas certainly is relevant to PNs, no concrete solutions to PN problems are proposed. Hence, we leave them out of Table 3.1, where we list the concrete approaches and how they address the various user requirements of Chapter 2. In the table, a 'Y' indicates that a particular project does consider topics related to that particular requirement to a significant degree. However, it does not necessarily mean that the project fully addresses that requirement. Indeed, none of the projects address all the requirements.

The projects that have been running in parallel with MAGNET, MAGNET Beyond, and PNP2008, that is PDE, MyNet, and PUCC, are the ones that come closest. Further, none of the projects cover context awareness. This is not to say that there are no context awareness projects. In fact, this is a very active area of research. A representative example of a large project on context awareness is the Dutch Freeband Awareness (http://awareness.freeband.nl/). MAGNET Beyond and PNP2008 specifically consider context awareness in personal networks.

Table 3.1 Summary of related projects and requirements.

	Ubiquitous Networking	Heterogeneous Hardware	QoS & Reliability	Naming & Service Mgmt.	Context Awareness	Security and Trust	Privacy	Usability
Ambient Networks	Y	Y	Y					
PACWOMAN	Y	Y	Y					
SHAMAN						Y	Y	Y
PDE	Y	Y	Y	Y		Y		Y
MyNet		Y		Y		Y		Y
PUCC				Y		Y	Y	
MOPED	Y	Y	Y					

A 'Y' in the table indicates that the project addresses the requirement, but does not necessarily completely meet it.

3.12 Summary

In this chapter, we have discussed research projects and technologies proposed in the area of personal and wireless communication, including work aimed at either analyzing future personal communication requirements or building integrated architectures and providing concrete solutions. We have also discussed major research areas, such as ad hoc networking and pervasive computing, from which personal networks originated.

We showed that very few attempts have been made to address all aspects of personal networks, leaving users with only fragmented and incompatible solutions. Hence, we conclude that more research and development is required to bring personal networks to fruition.

4

The Personal Network Architecture

In the previous chapters, we introduced the concept of personal networks and what we expect from it. We also showed that there is no integrated solution to all of the user requirements that we expect a personal network to support. In this chapter, we propose an architecture for personal networks that supports or can support all the requirements outlined in Chapter 2. By 'architecture', we mean a high level description of the entire system, its components, their relations, interfaces, and main functionalities. The purpose is to bring structure, separate the various concerns, and make it easier to describe the system. The personal network architecture is described at a high level, focusing on networking aspects. The main purpose is to give the reader a complete picture in order to better understand the remaining chapters and how the various solutions fit together.

It is important that the architecture does not violate any of the personal network user requirements or limit the possibility of fulfilling all requirements. The complete personal network architecture with its solutions must allow seamless communication among heterogeneous networks, work on mobile devices, offer reliable communication, be context-aware, provide security and privacy, and allow the complete system to be easy to use.

This chapter is organized as follows. We start with a few words on terminology in Section 4.1, then we introduce the core idea behind the PN architecture in Section 4.2. In Section 4.3, we divide the architecture into three levels and explain each of these levels. Section 4.4 covers how new devices are introduced into a PN. Networking aspects at a local level, that is, how personal devices can communicate directly with each other without external assistance, is discussed in Section 4.5. Section 4.6 deals with ways to connect distant personal nodes over infrastructure networks, while Section 4.7 covers communication with nodes belonging to other persons. Section 4.8 discusses higher layer support systems, such as service discovery and context information management for PNs, while Section 4.9 introduces the concept of federations of PNs. In Section 4.10, we discuss the ability of our proposed architecture to meet the requirements of PNs. Section 4.11 concludes this chapter with a summary.

Personal Networks: Wireless Networking for Personal Devices   Martin Jacobsson, Ignas Niemegeers and Sonia Heemstra de Groot
© 2010 John Wiley & Sons, Ltd

4.1   Terminology

When we define the PN architecture in this chapter, we will do so by introducing new terms and concepts. All the terms that we introduce in order to describe the architecture are listed and further defined in Appendix A as an aid to readers who require a more in-depth and precise understanding of the terms and concepts. However, all terms will also be introduced throughout the remainder of this book. When we introduce or refer to a term that is defined in Appendix A, we will indicate that by putting the term in italics.

It is important to point out that in this and later chapters the *personal network* is a technical concept that is different from our use of the term earlier in this book. Previously, the term 'personal network' denoted a vision of what we are trying to achieve; henceforth it will be a technical concept in pursuit of that vision.

4.2   Personal and Foreign Nodes

The main concept in this architecture revolves around the distinction between *personal nodes* and *foreign nodes*. Personal nodes become part of one's PN, while foreign nodes are kept outside. To make this happen, each *node* must know which PN it belongs to and must be able to detect whether another neighboring node is a personal node or not (i.e. belongs to the same PN). The process of introducing a node into a PN is called *personalization* and is a prerequisite to PN formation. Personalization should only take place once, when a new node is acquired for the first time by the user. A node then permanently remains a personal node until the user decides otherwise. Which nodes the user chooses to personalize is obviously up to the user, but it could include nodes he owns or otherwise possesses for an extended period of time. By extending the example that we introduced in Section 1.3, we can explain this concept as follows:

> *Jane has just bought a new digital camera. If the camera is part of Jane's PN, it can immediately store photos on Jane's home storage server, which has a bigger capacity and is more secure. Further, if Jane is carrying her navigation system, which is equipped with a GPS receiver, then the camera can access that information and geo-tag every photo she takes. To enable this, Jane needs to include the newly acquired camera in her PN. To do this, Jane places her mobile phone, which is already part of her PN, right next to the camera. She then switches the camera on. Using secure near field communication (NFC), the two devices discover each other and start to exchange information. Soon after, the mobile phone gets Jane to identify herself and then asks whether she really wants to include the camera in her PN. Jane confirms and after a few more exchanges over the NFC channel, the camera becomes part of the PN and appears as a personal node on the screen of the mobile phone.*

4.3   The Three Level Architecture View

Our proposed architecture for personal networks is described using three levels of abstraction in order to better understand, structure, and localize the problems that need to be solved. This is a first step in developing the protocol solutions. Figure 4.1 shows the three abstraction levels and how they relate to each other.

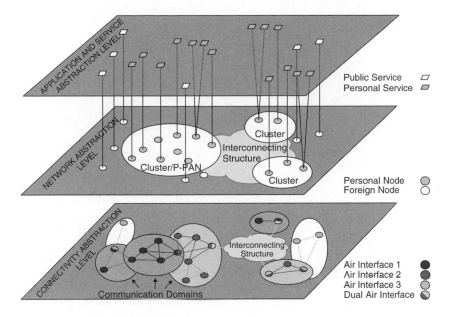

Figure 4.1   The three abstraction level view of a PN.

Proceeding from the bottom up, the first level is the *connectivity abstraction level*. Here, the devices can communicate with each other over common communication interfaces and medium access control (MAC) mechanisms. The *network abstraction level* is placed above the connectivity abstraction level. In this level, routing and other networking mechanisms reside, which enable communication among all personal nodes. To reflect the provision and usage of services in the PN concept, the *application and service abstraction level* is defined at the top. It contains all the applications and services offered by the nodes at the network abstraction level. Only the applications and services are in practice visible to the user. Also less obtrusive, but still important services such as name resolution and service discovery are part of this level.

When it comes to implementation, a protocol stack has to be chosen and adapted to fit the abstraction levels. All popular protocol stacks are layered, such as the OSI reference model's seven layers (ISO 1996) or TCP/IP's four layers (Stevens 1994). In any case, the mapping of the layers onto the abstraction levels will be a straightforward task once we understand the details of each of the abstraction levels.

4.3.1   Connectivity Abstraction Level

As mentioned earlier, PNs need to operate on as many as possible of the technologies at the connectivity abstraction level. Nevertheless, an accurate abstraction of this level is still needed in order to correctly design the higher abstraction levels and to understand the connections and interactions between the higher and the lower levels. This level is composed of devices with *communication interfaces* connected to each other via

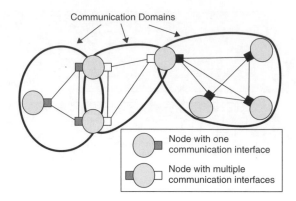

Figure 4.2 Nodes with communication interfaces communicating through communication domains.

different *communication domains* corresponding to given communication technologies. A Bluetooth piconet will be one communication domain. Each device in that piconet will have a communication interface attached to that communication domain. Figure 4.2 shows a more complex example where devices with several communication interfaces participate in several separate communication domains.

Problems addressed at this level concern the link layer, for example, medium access control, the physical link, and their interrelationships. Solutions for this abstraction level can rely on existing technologies, such as WLAN (IEEE 1999) and Bluetooth (IEEE 2005), but also future technologies, such as IEEE 802.15.3 (IEEE 2003). Since it is virtually impossible to adopt one single radio technology for PNs, we believe that many different kinds of radio technologies will be utilized. Some are suitable for short range high speed communication, while others are more suitable for access to the infrastructure. Furthermore, radio technologies will continue to evolve. Since older devices may use older technologies, it is important to enable gradual shifting from older technologies to newer ones.

At the same time, these different radio technologies may be so different that they will not be able to interoperate at the connectivity abstraction level (e.g. IEEE 802.15.1 (IEEE 2005) and 802.15.3 (IEEE 2003)). Therefore, we need devices with multiple radios and a network abstraction level that can glue all these different radio technologies, both current and future, together.

4.3.2 Network Abstraction Level

What is called a device at the connectivity abstraction level becomes a node at the network abstraction level. In most cases, it is sufficient to assume that node and device are the same. However, there are two exceptions. The first is devices without networking capabilities – they will never become nodes. The second is devices shared by two or more persons. Such devices can host two nodes, where one node belongs to one PN and the other node belongs to another PN. This concept allows devices to be shared by several persons. However, we will rarely consider shared devices and will frequently use the two terms interchangeably unless the distinction is important.

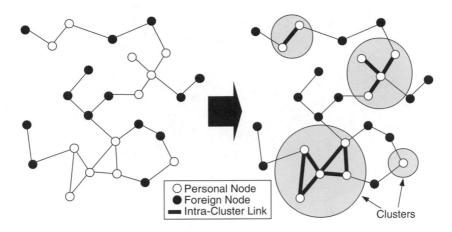

Figure 4.3   Cluster formation.

Given a single user, there are two types of nodes at the network level: personal nodes and foreign nodes. Personal nodes are nodes that belong to the user, while all other nodes are foreign nodes. The PN is the collection of all the user's personal nodes, regardless of their locations. This includes both nodes that are remote and nodes that are in the near vicinity of the user. Active personal nodes that can communicate with each other without external assistance are grouped into *clusters*. Typical PNs will consist of a few disjoint clusters located in areas where the person has his devices, such as one at home, one in the car, and one around him.

Figure 4.3 shows a scenario where some active nodes are scattered in an area. Some of the nodes are personal nodes, as can be seen in the left-hand part of the figure. The thin lines denote possible direct communication links. The cluster formation mechanisms of the personal nodes will detect when two personal nodes are neighbors (i.e. can communicate directly over a link, or, to use our terminology, over one single communication domain). If so, a secure connection is established over the link, denoted as thick lines in the right-hand part of the figure. Using these secured links, which we call *intra-cluster links*, connected subnetworks of active personal nodes are formed. These subnetworks are what we call clusters.

A user is likely to have several active clusters – a home cluster, an office cluster, a car cluster, etc. The cluster directly around the user was also called the *P-PAN* in the earlier PN literature and was assumed to consist of carried and wearable personal nodes. In other words, the PN is an extension of the P-PAN and may contain several active clusters, both far away from and in the vicinity of the user (see, for instance, Figure 4.4).

The network level architecture separates the communication among personal nodes of the same PN from the communication to, from, and among other nodes and devices. To do this, nodes need to be able to distinguish personal nodes from foreign nodes among their neighbors. This is the first step in cluster formation and should be fast and lightweight as well as secure against currently known attacks such as replay, man-in-the-middle, and denial-of-service (DoS) attacks. When two or more personal nodes find each other, they start to exchange routing information and other kinds of

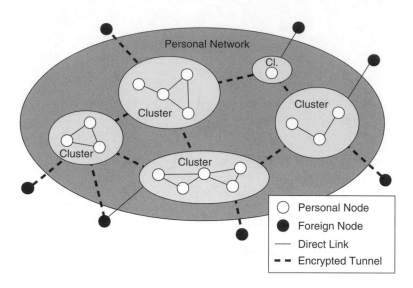

Figure 4.4 PN network level view.

networking information and thereby form a cluster consisting of only personal nodes. Foreign nodes are kept out of all this to protect the organization of the cluster. Section 4.5 discusses cluster organization further.

> *When Jane is on the move, she often brings her mobile phone, laptop, and camera. Using PN networking technology, these three nodes will automatically detect each other and start exchanging messages to form a cluster. The cluster consists only of Jane's three nodes, and the three nodes themselves handle all routing and communication within the cluster. Even if there are other people around Jane with their own PN nodes, Jane's cluster will consist only of her three nodes. This will ensure the network security of Jane's PN and obviously allow the nodes to share services and content.*

Not all the personal nodes in one PN will be able to communicate with each other in this way due to the characteristics and limitations of the radio technologies available to cluster nodes. Instead, a PN will consist of several clusters at various remote locations. The only way to connect the different clusters with each other is to use *interconnecting structures* (such as the Internet). To do this, tunnels between the clusters will be established and maintained. Section 4.6 has more to say about these inter-cluster tunnels and other aspects of PN organization.

> *Jane's mobile phone has a UMTS connection to the Internet. On behalf of the cluster of Jane's three nodes, the phone can establish an access link. Using this link, tunnels to clusters of Jane's nodes at home, in the office, or elsewhere can be established. Using these tunnels, all Jane's nodes can communicate, even while she is on the move. The PN networking functionality updates the tunnels as Jane moves about.*

When Jane's laptop can establish a high-speed link via a WLAN access point, her cluster can switch the tunnels to use the high-speed link instead. Further, when Jane arrives back home, the home cluster and the cluster of the three nodes merge into one single cluster and thus avoid using slow and expensive interconnecting structures.

Though a lot of communication will take place within the PN among a user's personal nodes, it is also important that communication can take place between PNs and with non-PN devices. We call this *foreign communication*, and cover it in Section 4.7.

4.3.3 Application and Service Abstraction Level

On top of the network abstraction level, we find the application and service abstraction level. This level consists of *applications*, *services*, and *clients* running on the nodes. Some nodes offer services to others, while some only use services offered by others. Whether the node where a service is running is a personal node or not also influences this abstraction level since it sits on top of the network level.

Services can be confined to the PN. This implies that these services can only be used by the person himself within the PN. This sort of access control does not always offer satisfactory protection. The real user of a personal node is not guaranteed to be the user of the PN and there is no way for the service to know who the real user is without additional authentication. Therefore, some very sensitive private services may still require, for instance, passwords or biometric authentication.

Services can also be offered to anyone the PN user wants to share the services with. A service can be offered both by someone's PN and by commercial providers. Some public services may not need any special trust, while others may still require establishment of ephemeral trust between the service node and the client node. In some cases, service usage will be charged.

To further improve the usability of a PN, a service discovery and management framework is defined that can support the applications and enable auto-configuration and adaptation. Figure 4.5 shows an example of how a PN service framework can work. In this case, we show the service framework proposed by IST MAGNET (http://magnet.aau.dk/). Each cluster has a *service management node* (SMN), which is elected among the cluster nodes and works as a repository where all services are registered. Clients in the cluster that wish to use a particular service can query this repository. To facilitate usage of services across clusters, the SMNs in the different clusters communicate with each other and share service information. For foreign communication, the SMNs can advertise the shared services of the PN to surrounding foreign nodes. The SMN may also register foreign shared services that it discovers in the vicinity of the cluster. This information can be used by a client on a personal node that wishes to use a foreign service. In addition to service discovery, the SMNs may also control the ongoing service sessions, that is, a client connected to and using a particular service. The SMN may, for instance, terminate or reconfigure ongoing service sessions when the network capacity has been exceeded.

Furthermore, applications and services can benefit from knowing more about the context they operate in as well as the context of the user. In order to support this, a context management facility is needed in a PN. This facility should be able to supply

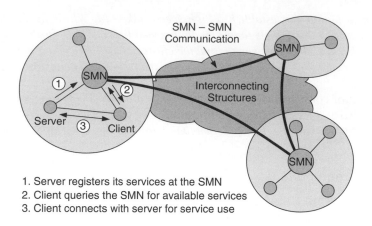

1. Server registers its services at the SMN
2. Client queries the SMN for available services
3. Client connects with server for service use

Figure 4.5   Service discovery in a PN according to IST MAGNET.

the relevant context information to any application or to system functions at other levels. Moreover, it should be able to collect and store this context information. Given the diversity of context, it should also be an extensible facility, able to incorporate any new type of context deemed useful.

Service discovery and context management as well as many other features related to applications and services are covered in more detail in Chapter 8.

4.3.4   Interaction between the Levels

Mechanisms at the various levels need to interact with each other in many different ways. The more information flows between the different levels, the better they can operate due to the extra knowledge of what is going on elsewhere. This is usually known as cross-layer information and includes information from the physical layer and all the way up to the application layer. However, not only cross-layer information within the node itself is useful, but also context information, coming from neighboring nodes or by sensing the environment, can sometimes substantially improve certain decisions. To better manage this information, a cross-layer and context information management framework can be used. We will discuss such a framework in Chapter 8.

4.3.5   Distribution of Networking Functionality in PNs

It is important to realize that the functionality can be distributed differently at the different levels. For instance, networking can be handled in a distributed way with minimal functionality placed in a central point. This may give robustness and efficiency at the network level. At the same time, service functionality can be centralized at the service abstraction level. For instance, the service discovery for PN-internal services can be done via a central service repository.

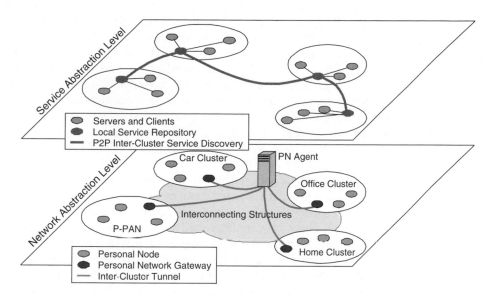

Figure 4.6 An example of functionality distribution in a PN.

Which solution is best depends on the abstraction level and may be independent of the solutions selected at the other levels. Figure 4.6 shows an example of a PN where the networking functionality in the clusters is distributed, but centralized between the clusters. At the service abstraction level, the approach is the opposite; service discovery is centralized within the cluster, but fully distributed (using peer-to-peer technology) for inter-cluster service discovery.

However, these are merely alternatives within our PN architecture. The design decisions depend on many factors, such as the type of applications considered, the interests of business stakeholders, the concerns of particular users, and, whether we are considering short-term solutions involving tens of devices per user or long-term solutions aiming for thousands of devices per person.

4.4 Personalization of Nodes

A core concept of a PN is the distinction between personal and foreign nodes. The personalization mechanism is responsible for establishing all the necessary configuration on the personal nodes and thereby enabling nodes to know which PN they belong to and to be able to tell whether a neighboring node is personal or foreign. For this, automatic and efficient mechanisms are needed. The obvious next question is how to make a node into a personal node.

To start, a dedicated *personalization device* or the first node a user buys must be manually configured by the user. The PN is given a name and perhaps a unique PN identifier to distinguish it from other PNs. Other settings can be configured at this time, such as the address of the PN agent (which will be described in Section 4.6). The next step is to personalize the rest of the nodes that the user possesses.

There are many different ways in which personalization can be done. It can, for instance, be based on subscriber identity module (SIM) cards, similar to the ones used in mobile phones today (3GPP 2005). Each user has a couple of SIM cards that share some common secrets. By inserting a SIM card into a device, it becomes a personal node and part of the user's PN. This is how 3GPP proposes to implement their version of 'personal networks' (3GPP 2009b). This approach requires the user to get these SIM cards from somewhere and all PN devices to be equipped with SIM readers. Another potential personalization process is based on the pairing of personal nodes.

In the pairing process, the node to be personalized establishes a secure connection using a shared communication domain to an already personalized node (or a dedicated personalization device). Using this connection, they exchange security keys and thus establish a permanent security association. These keys are later used to securely detect each other and to establish secure connections. At the same time, important configuration settings can be given to the newly personalized node. Depending on the scheme, the pairing can take place with any earlier personalized personal node or a dedicated personalization device. Chapter 9 will discuss these schemes in more detail.

The end result of the personalization of a device is that it is fully included in the PN. It becomes a personal node. As such, the new node gets access to the other nodes in the PN and grants them access to itself. To be precise, it gets and gives access to the PN networking mechanisms. Hence, the new node can generate and trust routing messages, hello messages, data packets, etc. Since only personal nodes can participate in this, only nodes that are under the user's control are included in these mechanisms. Therefore, all nodes in the PN can be assumed to be cooperative. In this way, many security-related problems with respect to networking are solved as we protect the entire network from unauthorized nodes. However, this requires the personalization steps to be secured by means of user authentication.

Finally, there is also the need to reverse the personalization of a personal node. Reasons may include a change of owner, a device being discarded, or when a personalized device is compromised. In any case, it must be possible to invalidate the personalization. Only the rightful owner should be able to do this, and it must involve authentication in the same way as personalization involves authentication.

4.5 Cluster Organization

At the connectivity abstraction level, nodes that have direct connectivity (a common communication domain) and belong to the same PN can establish secure communication. When more personal nodes are discovered, with which secure communication can be established, a cluster is formed. A cluster consists of only personal nodes and the secured communication links between them, as explained in Section 4.3.2. Once secure connectivity has been realized, communication at the network level can take place between all personal nodes in the cluster, without using foreign nodes.

The tasks of cluster formation are to use short range communication technologies to discover neighboring personal nodes, establish secure links between them, and maintain the routing. We propose to use all short range communication opportunities to find all neighboring personal nodes and thereby making the cluster as large as possible. To rely on short range communication technologies as much as possible is likely to be more efficient than using long range mobile networks, such as UMTS or LTE. Short range technologies do not have the same range as mobile networks that may cover a whole country or more. On the other hand, they are more likely to offer much higher bandwidth for the same energy consumption and interference (due to spatial reuse). At the same time, they have higher availability as they do not require infrastructure support in forms of networks and servers. All this makes them preferable for use in clusters when compared to infrastructure-based mobile networks (see Figure 1.1 in Chapter 1).

Clusters are dynamic in nature. Nodes are switched off or become available. They may also roam between clusters. Furthermore, clusters can split, for instance, when a person brings some nodes from an existing cluster with him and leaves others behind. Likewise, clusters can merge, for example when a person arrives home with his wearable personal nodes and they merge with the home cluster. Potentially, there is no limit to how large a cluster can grow, both in terms of number of nodes and geographical span. However, typically we expect clusters to have a small number of nodes and a limited geographical span, because of the way they will be deployed. Typical radio technologies used for intra-cluster communication will have limited range (e.g. IEEE 802.15.3 has a typical range of 10 meters), which means that normal clusters will be limited in size.

Not all nodes in a cluster will have direct links to all other nodes, because of different link layer technologies or radio range limitations. This implies that a cluster might be a multi-hop network. Therefore, it must provide addressing and routing functionality in order to enable efficient communication. This functionality should be able to deal with the specific characteristics and dynamics of clusters: nodes can roam, join, and leave the cluster. In order to handle these characteristics and dynamics, a distributed, totally self-organized, and efficient multi-hop mobile ad hoc routing mechanism should be used.

The cluster around the user is the P-PAN. The P-PAN operates with mechanisms similar to those of clusters. Due to the dynamic behavior of the clusters, personal nodes in the P-PAN or any of the clusters may roam and join other clusters or the P-PAN. It is therefore natural that all the clusters of a PN, including the P-PAN, use the same cluster organization mechanisms for better integration when nodes roam or clusters merge or split. Hence, there is no need to handle the P-PAN differently from the other clusters at this level. Because of this, we will use 'cluster' to denote any cluster or the P-PAN. Only at the application and service abstraction level it may make sense to distinguish the P-PAN from the other clusters, because the P-PAN is the cluster closest to the user.

The use of the Internet Protocol (IP) as a common base makes it possible to have a network layer architecture that is independent of the heterogeneity of the underlying link layers. Both IPv4 (Postel 1981) and IPv6 (Deering and Hinden 1998) are possible for the intra-PN traffic. However, since most parts of the Internet today use IPv4, it is important that the tunnels between the clusters support IPv4 networks.

Clusters can also consist of very limited devices that have no IP capabilities. Examples of such devices are networked sensors and simple actuators. As these devices can still offer important services to other nodes in the cluster and the rest of the PN, they should still be connected. Therefore, a personal node that can act as a bridge between the IP-incapable device(s) and the cluster should enable this. The bridge can make the services offered by the IP-incapable device(s) accessible to the rest of the cluster and the PN.

Finally, a cluster will not only operate as a standalone network, but also interact with its immediate environment, such as nearby foreign nodes or networks. Nodes in a cluster that can provide connectivity to nodes outside the cluster are called *gateway nodes*. Such nodes must have some special functions, such as address translation, filtering of incoming traffic, setup and maintenance of inter-cluster tunnels, etc. These tasks might be quite heavy for some devices, so it is useful to select powerful personal nodes as gateway nodes when possible. The process of finding capable gateway nodes with links to foreign nodes or networks is another network function that is provided by the clusters.

Chapter 5 will further discuss these and other issues related to clusters.

4.6   Personal Network Organization

A PN can have multiple clusters that are geographically dispersed. As stated earlier, each of these clusters uses the same mechanisms, but they organize themselves in a completely standalone and independent fashion. A PN realizes communication between these remote clusters using the trust already established by the personalization step. By 'remote' we mean that communication between the clusters can only be realized with help of foreign nodes and by using network layer routing and forwarding. Networks of foreign nodes, which can be used to interconnect clusters like this, will be referred to as *interconnecting structures*.

In order to form a PN, three requirements need to be fulfilled by the PN organization mechanism:

(i)   When access to an interconnecting structure is available, the clusters need to be capable of locating each other.
(ii) Once they have located each other, they must establish secure tunnels between themselves.
(iii) Last but not least, once the PN has been formed, it must maintain itself in view of the dynamic nature of the networks. That is, the tunnels must be updated when one or more clusters roam, such as when the car of a car cluster moves.

Gateway nodes are personal nodes in the clusters that have connectivity with either foreign nodes or the interconnecting structure. It is their responsibility to construct the interconnecting communication needed to connect the clusters of a PN. This includes looking for opportunities to establish such communication. In order for each cluster to locate the other clusters in the PN, the existence of an agent constitutes a big advantage. Such an agent is referred to as a *PN agent* and every PN should have one. Its role is to coordinate the clusters and keep their locations in a database. In this way, clusters within a PN can easily find each other. Figure 4.7 shows a PN with a PN agent assisting the clusters in connecting them with each other using inter-cluster tunnels between gateway nodes in the different clusters. The PN agent should be considered as a functional concept and not as a device, as there may exist many different solutions to implementing the PN agent concept, including distributed solutions.

For the communication between the clusters, we propose to use tunnels. The purpose of the tunnels is twofold:

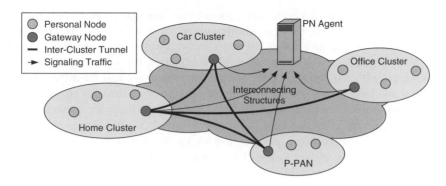

Figure 4.7 PN organization.

(i) They provide secure means for inter-cluster communication by shielding the intra-PN communication from the outside world.

(ii) They will be established and maintained dynamically to efficiently deal with cluster mobility. Further, they can also help connect clusters that reside behind network address translators (NATs) or firewalls.

Once the PN has been formed, intra-PN communication can take place. However, in order to establish connectivity among the personal nodes, addressing and routing are indispensable. One possible approach is to see the PN as a single large multi-hop ad hoc network in which most of the links are wireless, some are wired, and some are tunnels between the clusters. Within this ad hoc network, we can adopt a flat addressing scheme and run an ad hoc routing protocol that has been optimized for this environment. For instance, a PN-internal IP prefix could be reserved and all nodes within the PN will select a PN-unique IP address with this prefix. This IP address will be independent of the location of the node in the PN. This approach has the great benefit that, in combination with the dynamic tunneling mechanisms, mobility will become completely transparent for the higher layer protocols. The ad hoc routing protocols will hide intra-cluster mobility and the dynamic tunneling will hide cluster mobility and gateway node changes.

In Chapter 6, we discuss the PN organization further, including various PN agent solutions as well as some other alternative solutions for PN organization.

4.7 Foreign Communication

Hitherto, we have only covered communication between personal nodes. However, a PN cannot exist in isolation, but needs to interact with other PNs as well as PN-unaware nodes and other non-IP devices. Foreign communication involves both using services from foreign nodes and offering services to these nodes.

At the connectivity abstraction level, at least one personal node must obviously share a common communication domain with the foreign node to be able to establish any communication. If this is the case, then the network level must provide the PN with a way to also communicate with foreign nodes without compromising its own security or adversely influencing the intra-PN mechanisms.

As stated earlier, personal nodes that connect to a foreign node are called gateway nodes. Gateway nodes need to treat foreign traffic in a different way from intra-PN traffic. They must, for instance, block all non-approved traffic from entering the PN. Furthermore, gateway nodes must bridge all mechanisms used inside the PN with those used to communicate with the foreign nodes as these mechanisms will be different. If a personal node wants to communicate with a remote foreign node through an interconnecting structure, then the gateway node that links the cluster up to the interconnecting structure needs to act as the bridge. In this case, the gateway node bridges between the PN-internal mechanisms and the mechanisms used on that interconnecting structure.

When a foreign node wishes to establish communication with the PN, and when no direct local connection is available, it can turn to the PN agent of that PN via the interconnecting structure. Foreign nodes only need to remember the address of the PN agent of a PN to be able to initiate connections with that PN. To simplify even further, the address of the PN agent can be given a name that can be resolved through, for instance, DNS. Since the PN agent knows the location of all the clusters in its PN, it can tunnel the packets to the appropriate cluster and personal node. At the same time, the PN agent may also be used to bridge between the interconnecting structure mechanisms and the intra-PN mechanisms.

Chapter 7 contains more details about foreign communication.

4.8 Higher Layer Support Systems

Service discovery and context management are two examples of application support systems that greatly improve the user experience. With such systems, PN applications can more easily adapt to changing environments, be more proactive and more responsive to the user. With the introduction of the service concept, we are able to make a more reconfigurable system since services can be replaced on the fly. Having context awareness makes the system adaptive to the environment and the user. As a result, many more automatic tasks become available.

Several application support systems have been proposed, including many different ones for service discovery and context management. However, the special characteristics of PNs render many of them unsuitable. We require these systems to support poor connectivity or lack of connectivity, mobility, scarce resources, etc. Ideally, such systems will be aware of the structure and composition of PNs and able to use them to optimize their performance. At the same time, they need to support the requirements of the PN applications.

When Jane woke up this morning, she wanted to print an important document. However, she ran out of paper in her own printer at home and the printing job was canceled. When she arrives at her office, her PN discovers that there is an operational printer in the office. The application asks Jane whether she wants to resume the printing job with this printer. She clicks yes and picks up her document from the office printer a little later.

In Chapter 8, we discuss these application support systems for PNs in much more detail. Among many other things, we discuss service discovery, context information management, and content management.

4.9   Federations of Personal Networks

Human beings, by their very nature, are communicators. While PNs focus on the communication between nodes belonging to a single person, there are many scenarios where it is desirable to extend this communication to people and their services outside of the PN. The driving force behind these scenarios is the secure interaction of multiple people with a common interest in achieving a specific goal, or set of goals. Examples of such scenarios include conferences, research projects, virtual meetings, family networks, inter-vehicle networks, emergency networks, distance learning networks, and commercial/private resource sharing services. Groups involved in these common interest examples could include family members, colleagues, emergency teams, friends, students, online gamers, etc.

For this purpose, we define the concept of a *personal network federation* (PNF). We introduce it with an example:

To strengthen their knowledge, Jane's boss decides to send Jane and three of her colleagues to a conference. After they return from the conference, they all want to look at and share the photos and videos that they took. For this purpose, they form a PNF that contains a subset of their PNs. This subset is composed of a camera phone and a laptop (PN1), a computer with photo editing software (PN2), a digital photo camera and a photo printer (PN3), and a digital video camera (PN4). All this is illustrated in Figure 4.8.

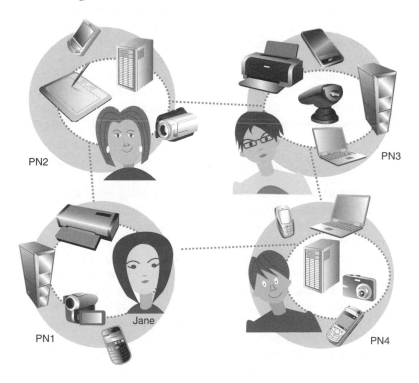

Figure 4.8   A PNF composed of four PNs to share photos and videos.

In this PNF, the four colleagues can share with each other several types of content
(e.g. photos and videos) and services (e.g. display, photo-editing, and printing ser-
vices). Note that because the PNs only share a subset of their resources, those PN
resources that are not included in the PNF will not be visible or accessible to the
other PNs. Once the four colleagues have finished sharing the videos and editing
and printing the photos, they dissolve the PNF.

Chapter 10 is devoted to further discussion of PNFs.

4.10   Discussion

Now that we have introduced our architecture for PNs, we should ask ourselves two
questions:

 (i) Can this architecture be implemented in reality?
(ii) Does it then fulfill the requirements outlined in Chapter 2?

   The first question will be answered in the following chapters where each part of the architecture will be further discussed and developed to a level where the architecture becomes
concrete. However, answering the second question is actually more important, since if
the architecture does not meet the requirements, then all the development is for nothing.
   It is hard to know in advance whether the requirements can be met by the architecture
since it depends on the detailed solutions selected for each part of the architecture. For
instance, QoS and reliability depend heavily on the used routing protocols and mobility
mechanisms, which are not dictated by the architecture. Therefore, we cannot say whether
the requirements have been met. However, it is important that the architecture does not
prohibit us from meeting the requirements. In the remainder of this chapter, we will try
to answer the question by discussing some known issues and their potential solutions.

4.10.1   Why a Network Layer Overlay?

The first question one must ask about the architecture is: Why define the PN at the network
level and in effect create a network overlay? An option would indeed be to define the PN
at the service level, by perhaps building a service overlay (or a peer-to-peer network) on
top of existing network solutions. While this certainly is possible, it would leave many
important issues unsolved at the network level. Users would still need to configure many
things, assign network addresses, routing protocols and other network settings for all kinds
of different networks. Current auto-configuration solutions are either targeted toward only
one network type (such as a home network) or do not handle all aspects of PNs (e.g.
cannot handle mobility). It is also likely that different types of devices use different types
of incompatible network solutions that would make integration unnecessarily hard.
   In the PN architecture that we defined, a unified network solution is proposed that
enables automatic configuration and adaptation. It provides connectivity between personal
nodes and is protected from uncooperative nodes. Also foreign communication is possible.

4.10.2 How Protected is a PN?

Personalization is the main concept of this architecture. It is used to make the distinction between personal and foreign nodes. Only nodes that have been personalized (i.e. personal nodes) can be part of the user's PN. In this way, the intra-PN network mechanisms within a PN can be protected from non-trusted foreign nodes.

When a node is being personalized, it establishes long-term trust with all the other already personalized nodes in the same PN by means of cryptographic methods. This actually means:

 (i) the new node gets access credentials from the other nodes in the PN;
(ii) the new node gives access credentials to the other nodes in the PN.

The next question to be answered is of course: what exactly does this access include?

One essential capability is that it gets and gives access to the intra-PN networking mechanisms. The node can generate routing messages, hello messages, data packets, and so forth that are trusted by its neighboring personal nodes as well as verify the authenticity of such messages it receives. Since only personal nodes can participate in this, only nodes under the control of the user are included in these mechanisms. Hence, the nodes can be assumed to be cooperative and in this way, many security-related problems with respect to networking can be avoided. Personal nodes can send packets among themselves and be sure that packets they receive originated from a personal node. Packets from foreign nodes are either filtered or are treated in such a way that an end node can distinguish them from packets coming from personal nodes.

Another capability that a personal node gets and gives access to is services. That is, a personal node allows access to its services from any other personal node belonging to the same PN. The benefit of this is obvious. This access control policy is very trivial, which hopefully means that most users can easily understand it. On the other hand, it has serious security drawbacks. If this policy is always used, then anyone can use a personal node to access any service in the whole PN and this can happen since personalization can only authenticate nodes, not the user of a node.

This problem includes lost or stolen personal nodes as well as nodes temporarily in the hands of someone else. If such a node is capable enough, it can be used by anyone to access any service within its PN, including sensitive services such as private photo albums, banking services, etc. The owner can, of course, exclude lost or stolen nodes from the PN. However, to only depend on a user action is not necessarily enough, since it may already be too late. Nevertheless, for many personal services, this weak level of security may still be acceptable, but since personal nodes certainly will be lost or stolen, there is a need for more security for more sensitive services.

A potential solution is to make sure that each personal node authenticates its direct physical user before the node can be used to access services on other personal nodes. However, then the security data used to implement the personalization and stored on each personal node needs to be properly protected. This requires good user authentication and good tamper-resistance on all personal nodes. Unfortunately, good tamper-resistance for every personal node is a very difficult problem and currently not feasible (Stajano 2002b). It is expensive and makes devices bigger, clumsier, and may still not yield good enough security.

As a consequence, to allow a personal node access without authenticating the user is not secure enough for sensitive services. For such services, additional authentication and access control on the service node itself are required. Hence, it must be possible for the user to add security measures when he feels the need for it.

4.10.3   How Usable is the PN Security?

One of the big concerns with the architecture is its security system and whether the users can handle it properly. Many other security systems, such as Pretty Good Privacy (PGP) (Whitten and Tygar 1999), have failed simply because users do not understand them. As before, it is impossible to answer this question without first implementing the system and conducting a usability study. Nevertheless, we can make sure that the architecture follows design principles known to work well, such as the ones outlined in (Yee 2002).

The design principles regarding usable security identified by Yee can be summarized as follows: explicit authorization, visibility, revocability, path of least resistance, expected ability, appropriate boundaries, expressiveness, clarity, identifiability, and trusted path. Some of these principles, such as path of least resistance, identifiability, and trusted path, have more to do with the user interface than the system architecture itself.

We believe that the other principles either are supported or can be supported by the final implementation. Personal nodes are manually included in the PN by the user (explicit authorization) and can again be excluded (revocability). Due to the explicit step of personalization, it should not be difficult to create a tool that displays this information and tells the user exactly which nodes are currently part of the PN (visibility). The remaining principles (expected ability, appropriate boundaries, expressiveness, and clarity) have to do with what the security system is capable of and how this is communicated to the user.

The distinction between personal nodes and foreign nodes can be based on, for instance, ownership, which is a concept people understand. The system is, at the same time, visible. Hence, the user can easily understand how it works and therefore create an accurate mental model of the system more easily (Norman 1988). With the help of the mental model, abilities and boundaries become clear and the user will be able to make the right decisions.

4.10.4   Do We Need to Manage Our PNs?

Today, private persons have problems updating and securing their own home PC. Considerable dissatisfaction is caused by having to install, update, and maintain sophisticated devices that are heavily software based, such as digital home theater systems, hard disk recorders, and digital cable or satellite tuners. What will happen when they have tens or maybe hundreds of wireless devices to maintain?

The idea is that PNs are self-configured and hence minimize the need for manual management. For many of the management tasks required today, it is clear that further technical solutions are required. A PN must automate as many of these tasks as possible in order to minimize the demand on the user to carry them out manually.

Automatic software updates and frequent backups of personal nodes are examples of tasks that a PN management tool must provide. This should not be a problem with the current architecture. These types of tools need to communicate with all personal nodes and perhaps a few services through the interconnecting structures, something that is already supported. With these connection possibilities, automatic solutions can be built as well as remote management solutions if necessary. Further, the system should inform the user when manual actions are required, such as switching on a deactivated node when there is a need for a software update. However, there is a need for a streamlined and common framework for this, considering that a PN may consist of several very different types of devices from different manufacturers with very different types of software.

4.10.5   What About the Social Dimension?

The real challenge with personal networking is actually to understand the social aspects, and here it must be agreed that the PN architecture is not fully sufficient. For instance, its focus on the person and her personal nodes makes it difficult to share devices. Most people have affiliations with a lot of different communities where devices actually are shared. A person might belong to a family, a company, a sports club, as well as informal groups of friends and acquaintances. The architecture must take these factors into account and properly support individuals in their social life. For instance, borrowing and sharing of personal nodes must be supported, perhaps only for a limited time in some cases, but also indefinitely. An example of the latter is a family sharing home equipment and appliances.

Occasional and exclusive borrowing of a device can be done by removing all sensitive information and giving the borrower full control of it. That is, all stored data is moved to another personal node and the personalization is temporarily deleted during the time someone borrows it. The node can be excluded from the owner's PN and temporarily included in the borrower's PN. Renting is a special form of occasional time-limited borrowing where this scheme also could be used. As long as there is only one concurrent user and the user does not change too often, this will work.

Further, our concepts of foreign communication and PNFs solve many of these issues. It becomes possible to interact with others through your PN. Communication can be established and information exchanged.

However, another solution is required when it comes to sharing a single node among several persons at the same time. Relying on foreign communication or PNFs is of course possible, but not always efficient. A shared device that is frequently used by two persons will never be part of both PNs and this may lead to suboptimal networking for at least one of the persons. As mentioned earlier, an alternative is to allow a device to be personalized by two or more persons simultaneously, but still keep the various PNs separate. A device with several personalizations need to be able to distinguish among the PNs and then maintain several concurrent but separate intra-PN networking mechanisms. This may be a bit heavy for mobile devices, but is not a problem for stationary and mains-powered devices that are more likely to be shared.

Sharing networks can also be beneficial. Consider, for instance, a home network belonging to a family. Such a network consists of several devices and network links that should be shared by all family members. It is not efficient if each of the family members has their own totally separate home cluster. It would be more efficient if their clusters would integrate into one family home cluster. After all, nodes belonging to family members are probably cooperative. It may therefore be beneficial if some PNs can be fully integrated at the networking level. Two PN users can, for instance, declare each other's PNs as totally trusted. Whenever a personal node meets a foreign node belonging to a fully trusted PN, it considers that node as a personal node, at least at the network level. At the application and service abstraction level, the node can still be considered as a foreign node to make sure that services are not accidentally opened up.

One more important social aspect is the conflict of interests between a person's professional and private life. On the one hand, companies set up security regulations on how an employee can use company devices and how to handle sensitive company data. On the other hand, that person has a private life in which he would like to use the same devices. This may include having a combined agenda for both professional and private activities. It may be desirable to be able to take care of some private matters when in the office or using privately owned equipment when working from home, etc. However, this will undoubtedly go against most company's security regulations if PNs cannot prove to be secure enough. Companies need to trust both the PN architecture and their employees' ability to use the system in a secure way before they allow their employees to have one PN containing both private and business devices.

4.10.6 More Issues?

As has been demonstrated here, the base architecture outlined earlier in this chapter is far from perfect. It tries to single out the trustable nodes from non-trustable nodes, but sometimes fails to do so. Personal nodes can be lost and therefore cannot always be fully trusted, while at the same time many foreign nodes actually are trustable and can be fully trusted. The architecture is very simple and easy to understand, but not yet sufficient. With this, we hope to have highlighted that these issues can indeed be solved. The current strength of this architecture is its simplicity. However, extensions are not only possible, but also required even if they make the PN more complicated. It is important to carefully consider which extensions are useful so that we can avoid creating an unnecessarily complicated system. One extension to PNs that we believe strongly in and will describe extensively in Chapter 10 is PNFs.

4.11 Summary

In this chapter, we have proposed an architecture for personal networks in which there are two types of nodes: personal nodes and foreign nodes. Personal nodes are nodes that belong to the user, while all other nodes are foreign nodes. The PN of a user is the collection of all that user's personal nodes. When active personal nodes come together and can communicate with each other, they form clusters. Personal clusters communicate with each other over interconnecting structures and thereby form a connected PN. The architecture separates the communication among personal nodes of the same PN from the communication to, from, and among other nodes. Hence, each node must know which PN it belongs to and must be able to know if another neighboring node is a personal node or not. This is achieved by a manual personalization step in which the user admits a newly acquired node into the PN.

To better introduce the various concepts of the architecture, we divided it into three abstraction levels: the connectivity abstraction level, network abstraction level, and the application and service abstraction level. At the connectivity abstraction level, we discussed link layer technologies, medium access, and other one-hop communication mechanisms. At the network abstraction level, personalization, cluster formation, PN organization, routing, and foreign communication are taking place. At the application and service abstraction level, we find the PN applications and services. At this level, we also include service discovery, context management, content management, and PN management.

In the final part of this chapter, we discussed the suitability of this architecture. We focused on the requirements introduced in Chapter 2 and verified whether the architecture could fulfill them all. Certain areas, such as security, usability, and social aspects, were analyzed in more detail. It was concluded that most requirements were fulfilled or could be fulfilled by our architecture. However, due to its simplicity, the architecture is not fully sufficient; some extensions are required, such as PN federations.

5

Cluster Formation and Routing

Now that we have introduced the overall architecture for PNs, it is time to look at the various parts of it. In this chapter, we start by looking at the clusters in more detail and their formation and maintenance in particular.

In Chapter 2, we argued that personal networks need ubiquitous network connectivity that is reliable. To achieve this, it is best if the personal nodes can manage their own communication among themselves without the need for infrastructure support. If there is no need for special communication equipment, such as access points and network switches, this will maximize the capability of ubiquitous network connectivity.

Each personal node should contain the necessary functionality to form and maintain the networks needed for its operation. Hence, when personal nodes come together, they should form a local ad hoc network that enables them to communicate with each other independently of external support. They should form a self-organized cluster.

This chapter is structured as follows. Section 5.1 precisely defines what a cluster is and introduces important requirements for cluster formation. Section 5.2 discusses mobile ad hoc networks, which is a technology that we propose to use to implement cluster formation and maintenance. Section 5.3 introduces the cluster formation mechanism and its neighbor discovery mechanism. In Section 5.4, we discuss unicast and broadcast routing in clusters. Section 5.5 summarizes the chapter.

5.1 What is a Cluster?

A cluster is a connected network consisting exclusively of personal nodes located within a limited geographical area, such as a house or a car. The nodes trust each other as defined by the personalization and are connected by one or more link layer technologies (MAGNET 2005a). All personal nodes should implement the same mechanisms at the network level so that they can easily find each other and communicate directly when there is connectivity between them at the connectivity level. In this way, the migration of a personal node from one personal cluster to another becomes easier.

A cluster facilitates secure network communication among personal nodes on a local scale using local communication technologies such as WLAN, Bluetooth, and Ethernet. It sets a boundary for the intra-cluster network level formation and maintenance mechanisms.

Personal Networks: Wireless Networking for Personal Devices   Martin Jacobsson, Ignas Niemegeers and Sonia Heemstra de Groot
© 2010 John Wiley & Sons, Ltd

A precise definition of a cluster is needed in order to clearly set a boundary between local and global network mechanisms. We therefore define a cluster as follows:

Definition: A *cluster* is a connected network consisting exclusively of active personal nodes and their communication domains.

Two personal nodes are in the same cluster if they can communicate with each other using a path between them consisting of only personal nodes such that each personal node shares a communication domain with the next personal node in the path. Hence, a cluster consists of personal nodes and the communication domains that connect them. A single personal node with no other personal nodes in its communication range is by itself a cluster, a single node cluster.

Two things determine a cluster. First, connectivity: do the nodes share common communication domains that connect them? Second, are the nodes part of the same PN? Figure 5.1 shows an example where several personal and foreign nodes are scattered in a small area (see also Figure 4.3). The personal nodes will find each other, establish secure links, and form clusters. Due to radio range limitations, it is impossible to form one cluster containing all personal nodes. Hence, in this scenario, two clusters are formed, and to connect the two we depend on an interconnecting structure, which we cover in the next chapter.

The communication technologies that are expected in a typical cluster are technologies that do not depend on deployed infrastructure. Usually, we envisage link layer technologies, such as local area networks (LANs), wireless local area networks (WLANs), and wireless personal area networks (WPANs). Such technologies have a limited range, but can be used ad hoc and usually offer high data rates. As a consequence, personal nodes within the same cluster are usually located within a limited geographical area, such as a house or a car.

Long range infrastructure technologies, such as UMTS, cannot be used within a cluster. Such technologies rely on several nodes not under the control of the cluster owner and

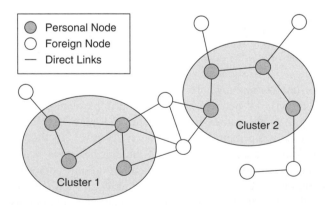

Figure 5.1   Example of two clusters.

can never be characterized as a single communication domain with one common MAC mechanism. Instead, they can be used when a person's clusters need to communicate with each other.

It is important to note that two neighboring nodes within a cluster belong to the same PN. Each node must validate whether its neighbor is a personal node or not. This is done by cryptographic means based on the pre-established security keys installed at the personalization step (see Chapter 9).

The purpose of defining a cluster in this way is to be able to use intra-cluster mechanisms to provide communication between as many personal nodes as possible, since local communication is likely to be more efficient than using global interconnecting structures. Interconnecting structures cannot always be assumed to be available and, in many cases, provide lower data rates and at higher cost than direct multi-hop intra-cluster communication.

There are, of course, alternative ways to define a cluster, such as using geographical distance. For instance, a cluster could be defined as personal nodes within a distance of 10 meters. This definition requires positioning capabilities and may not be suitable for network layer mechanisms even if approximating techniques are used (e.g. based on distance measuring using received radio signal strength). Geographical proximity among nodes within a cluster can, for example, be a good indication for security. Two nodes in the same cluster will most likely reside in the same location, such as inside a house, a car, or on the body of the PN user. Indeed, security based on this kind of information may be useful. However, it is unreliable and we should instead use mechanisms at other layers, which may provide more accurate context information. Another possible cluster definition could be based on a limit on the number of hops from an elected master node or a special device (Sulaiman et al. 2005).

Cluster formation and maintenance at the network level will take place in the same way in every cluster, including the cluster around the user, the P-PAN, and should therefore comply with the same set of requirements. We see the following requirements as crucial:

(i) The cluster formation and maintenance procedures must be as independent as possible of the lower layers, including MAC and link layers. At least current link layer technologies must be supported, including Ethernet (IEEE 802.3, 802.11, and 802.15), Bluetooth (versions 1.2 and 2.0), and future link layer technologies as far as can be anticipated.

(ii) A cluster can be connected using various link layer technologies. The cluster formation and maintenance functions must support multiple different link layer technologies in a single cluster at the same time.

(iii) The cluster formation and maintenance must be self-organized. It must be able to form and maintain itself without support from the user or external infrastructure-based equipment or services.

(iv) Personal node discovery and departure detection should be provided by the link layer. However, when no such functionality is available or it is inadequate, the network level has to provide this functionality. For improved quality of cluster networking, it is important that links between personal nodes are constantly monitored so that link breaks and node departures can be detected in an accurate and timely manner.

(v) Clusters may consist of very large numbers of personal nodes. It is then important that the cluster organization mechanisms are not detrimentally affected. Scalability is mainly an issue in intra-cluster routing.

Scalability may not be a very important requirement for the near future. In the beginning, mobile clusters typically will not consist of very large numbers of personal nodes. Perhaps a user will carry up to five mobile devices and have ten wearable devices. Non-mobile clusters, such as a home cluster, may contain more nodes. However, we expect that, in the future, very large numbers of devices will be connected and then scalability will become an important requirement.

For large numbers of sensors, special solutions, such as wireless sensor networking, will be required. In that case, the sensors themselves will never take part in a cluster. Instead, they will be 'hidden' behind a single personal node acting as a gateway between the PN and the sensor network.

Jane's camera only supports Bluetooth, while her laptop only supports WLAN. However, this is not a problem since her mobile phone supports both and the PN functionality will arrange the rest. Jane is in fact unaware of the fact that the three devices form a cluster and that a self-organized ad hoc routing protocol is doing a lot of the work. For Jane, cluster communication 'just works'.

Furthermore, the cluster detects new personal devices and selects the fastest communication links. This is something Jane is grateful for, especially when she is at home and wants to work with very large data files, such as video files. Data transfer is very fast, even though she does not have to configure anything.

In the remaining sections of this chapter, we will introduce and discuss options and solutions that can successfully meet all these requirements.

5.2   Mobile Ad Hoc Network Technologies

A cluster is basically a special type of ad hoc network. Ad hoc networks and their mobile variant, MANETs, were described earlier in Section 3.2. As mentioned there, MANETs are unplanned networks of mobile devices that can form quickly and automatically when the opportunity arises. As such, clusters will build on ad hoc and MANET technology.

In addition to routing, the Internet Engineering Task Force (IETF) MANET working group (http://www.ietf.org/html.charters/manet-charter.html) has recently started to focus on other subjects. New topics include a common packet format for signaling traffic in MANETs (Clausen et al. 2009c) and a neighbor discovery protocol called MANET Neighborhood Discovery Protocol (NHDP) (Clausen et al. 2009a). The new common packet format is extensible and has been extended for carrying the signaling of future ad hoc routing protocols, such as OLSRv2 (Clausen et al. 2009b) and DYMO (Chakeres and Perkins 2009). In the same way, it is possible to extend the packet format to also carry cluster-related information, such as cryptographic information related to the node personalization. However, one aspect not covered by the packet format is privacy. While it is possible to encrypt some signaling traffic, such as routing traffic, it is not possible for every signaling packet. Routing traffic is exchanged between personal nodes only after discovery and when a secure link has already been established. Hence, these packets can be completely encrypted by the link layer and there is no problem with using the common packet format in combination with any ad hoc routing protocol. On the other hand, node discovery packets and personal node authentication packets cannot be encrypted in the same way, since they are transmitted before a link has been secured. Therefore, they require special mechanisms to both securely authenticate the nodes and at the same time protect the privacy and the identity of the nodes.

NHDP, which of course is based on the common packet format, uses periodic hello messages. Each node periodically transmits hello messages on all its MANET-enabled interfaces. The hello messages contain the node's address, a list of its neighbors, and some other basic information. The purpose of the hello messages is twofold. First, they are used to discover new neighbors. Whenever a hello message is received on an interface from a previously unknown neighbor, a new link can be established. The second purpose is to monitor existing links and detect when they disappear. However, as explained in Section 5.4.2, this raises several issues not addressed by NHDP. For instance, using hello messages provides a very inaccurate view of a link's quality and is a slow way to detect changes.

The Ananas approach (Chelius and Fleury 2002) tackled the MANET problem by introducing an adaptation layer between the link layer and the network layer to emulate a normal switched Ethernet for the network and higher layers. This 'layer 2.5' actually uses ad hoc networking techniques to implement unicast and broadcast packet forwarding on top of a multi-hop network. The benefit of such an adaptation layer is that higher layers can continue to operate as if they were running on a normal fixed network. The drawback, however, is that too much information is hidden, which leads to inefficient and non-optimal operation. Protocols designed for switched Ethernet environments do not work particularly well in ad hoc networks as they are not able to cope with the special requirements that an ad hoc network poses, such as varying link quality and lower available bandwidth. However, the main problem is actually the dependency on broadcasting. Broadcasting in a switched Ethernet is both a reliable and relatively inexpensive operation. In an ad hoc network, that is not the case. Therefore, the Ananas model is not robust and efficient enough for clusters.

The idea of introducing a new layer for ad hoc networking between the link and network layers has been proposed by many others for other reasons as well. One popular reason is to provide a single abstraction of a link. Generic Link Layer (GLL) (Sachs 2003; Sachs et al. 2004) and Universal Convergence Layer (UCL) (MAGNET 2005d; Sanchez et al. 2005) are both variants of that approach. The idea is to avoid needing to develop the network layer to work with a particular type of link layer or to accommodate all the details of all the different types of link layer technologies. The reasons are that, on the one hand, a network layer needs to be independent from the link layer to allow it the flexibility to use any type of link layer, and, on the other hand, ad hoc networking requires a good amount of cross-layer interaction (Zhou et al. 2007a, b, 2009). A generic or universal abstraction of an ad hoc link by an adaptation layer can offer a good compromise between the two. The cluster formation mechanism can then use the same approach for several different types of link layer technologies.

5.3   Cluster Formation and Maintenance

The method we propose for cluster network formation is opportunistic in the way it forms a cluster. It seizes all opportunities and makes the cluster as large as possible using available ad hoc link layer technologies. These link layer technologies do not have the same range as mobile networks that may cover a whole country or more. On the other hand, they have much higher bandwidth for the same energy consumption. This makes them preferable for use in clusters when compared to infrastructure-based mobile networks. The cluster formation can be thought of as a network sub-layer function that makes use of available link layer technologies to form a cluster whenever possible. It is also proactive, meaning that it constantly looks for personal nodes in the neighborhood and possible links to extend the cluster. The purpose is to be able to use intra-cluster mechanism to provide communication between as many personal nodes as possible, since local communication is likely to be more efficient than using global interconnecting structures. Interconnecting structures cannot be assumed to always be available and, in many cases, provide lower data rates and at higher cost than direct multi-hop intra-cluster communication.

After a cluster has been formed, each personal node will have a list of direct neighbor nodes that are personal nodes. Associated with each neighboring personal node in this list are also security keys, which are used to protect the data messages on the link between the two personal nodes. The security keys are derived based on the personalization step as described in Chapter 9. Traffic from neighboring foreign nodes must be ignored unless the node is acting as a gateway node. Gateway nodes treat traffic from foreign nodes differently with firewall rules and the like. Properly encrypted traffic from neighboring personal nodes, on the other hand, is always trusted and, in this way, a secure cluster is formed. A cluster carries traffic among its personal nodes in a secure way and detects and filters traffic from foreign nodes. On top of these secured links and filtered neighbor lists, we introduce an intra-cluster networking layer with addressing and routing. Note that these functions will also be protected in the same way as any traffic between neighboring personal nodes. Hence, routing packets will also only be exchanged between the personal nodes.

5.3.1   Multi-Hop Clusters

Each of the steps to form and maintain a cluster is distributed. This is because we do not allow the cluster formation to depend on special nodes being present. Any two personal nodes that implement a common wireless technology must be able to find each other and form a secure cluster by themselves. Furthermore, distributed solutions are more robust, especially in multi-hop networks, and clusters may be multi-hop. That is, a cluster may not always be fully connected with direct links between all of its nodes. Even though clusters will be fully connected in many cases, we cannot exclude the possibility that a cluster is multi-hop and hence the cluster formation and maintenance procedures need to support multi-hop clusters. In particular, there are three reasons that may make a cluster multi-hop:

(i) Some link layer technologies have limitations on the number of devices in one PAN. The best example is Bluetooth (IEEE 2005), which can only accommodate eight active devices. If more active devices are present, a multi-hop scatternet (McDermott-Wells 2004) is needed in order to accommodate additional devices.

(ii) The radio range may not be enough to cover all devices in one hop. Imagine a house where a person's devices are spread all over that house, including home appliances. The radio range of typical WLAN or WPAN technologies in typical indoor scenarios will have problems covering a large house. In principle, a wired WLAN infrastructure with several access points could be deployed in such a situation, but there is actually no need for that. It is also possible to connect the devices in a multi-hop ad hoc network. If necessary, one can place mains-powered relay devices just to make sure that the multi-hop cluster is connected and covers the whole house.

(iii) It is necessary to accommodate several link layer technologies at the same time, for instance the different variants of IEEE 802.15 together with 802.11. If two nodes want to communicate but do not share a common radio technology, then a third device that implements both can act as a bridge; this is in fact a kind of multi-hop network. Obviously, it would be easier for the user to have only one single link layer technology, but in times of technology shift this will make a gradual change of technology possible. Another reason for multiple radio technologies in one PN is their diverse characteristics. Some technologies offer high data rates while others offer low power consumption, and unfortunately no single link layer technology satisfies all applications and requirements.

This means that we need to cater for multi-hop clusters and hence need multi-hop routing. However, the routing protocol should be optimized for small clusters that are fully or almost fully connected.

5.3.2   Link Layer Device Discovery

The first step toward forming a cluster is to detect neighboring nodes and their hardware and IP addresses. There are three different ways this could be achieved:

(i) Use the device discovery mechanism provided by the link layers. Several link layer technologies implement device discovery mechanisms. IEEE 802.11 uses beacons carrying the MAC address to detect neighboring nodes and access points. Bluetooth transmits inquiries to detect neighbors and form piconets. These messages can also be used by the cluster formation to detect new neighbors or the disappearance of neighbors. The advantage is that no new signaling is needed. However, different link layer technologies use different device discovery mechanisms and it is therefore necessary to define how new devices are detected for each possible link layer technology. Furthermore, it is possible that some link layer technologies do not have any device discovery mechanism that can be used by the cluster formation mechanism. Wired Ethernet is an example of that. In that case, other device discovery mechanisms are needed.

(ii) Use the neighbor discovery (Narten et al. 2007) of IPv6 (Deering and Hinden 1998). Unsolicited neighbor advertisement messages may be used to advertise a node's presence. This mechanism is mainly used to map IPv6 addresses to hardware addresses, but may also be used for detecting new neighbors. However, every node needs to transmit a neighbor advertisement periodically and that adds signaling traffic. IPv4 does not have such a mechanism.

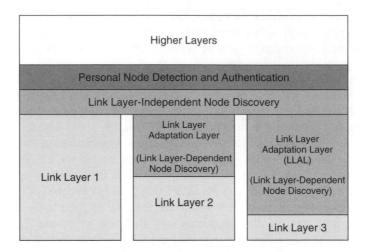

Figure 5.2 Link layer adaptation possibilities.

(iii) Implement a higher layer device detection mechanism. Hello messages based on the user datagram protocol (UDP) (Postel 1980) sent to an IPv6 link local multicast address would be a good option. By defining a new multicast address for IPv6 nodes that understands PN mechanisms, it is possible to filter out non-PN-capable nodes. These messages could also make the mapping between the IPv6 address and the hardware address just like the IPv6 neighbor discovery messages above. A similar approach is possible for IPv4 as well using either broadcast or multicast addresses.

The best solution would be to use the link layer device discovery when possible and otherwise use a higher layer mechanism. The link layer discovery option does not usually offer the mapping between hardware addresses and the IP addresses. However, this is not necessarily a problem since the personal node authentication will take place immediately afterward and the mapping can be part of that procedure instead. Figure 5.2 shows how a personal node that has different link layer technologies with different amounts of support could work. The leftmost communication interface (link layer 1) uses a link layer technology that has a discovery mechanism that can be used entirely by the cluster formation and maintenance procedure. The middle communication interface (link layer 2) has a discovery mechanism that needs only a small interface-specific link layer adaptation layer (LLAL). The interface to the right (link layer 3), however, may not provide such a discovery mechanism at all and hence its interface-specific adaptation layer needs to implement that.

5.3.3 Discovery of Node Arrivals and Departures

When a cluster node detects a new node on one of its communication interfaces, it will try to determine whether the new node is a personal node in the same PN or not. If it is a personal node, the cluster node will add the information of the new neighboring personal node to its list of known neighboring personal nodes and perform the necessary security mechanisms to establish a secure connection with the new node. The latter is covered in Section 9.2.

If the new personal node was not already part of the cluster, more network layer actions must take place:

(i)   Addressing issues might need to be resolved (see Chapter 6).
(ii) The routing protocol needs to update its topology (see Section 5.4).
(iii) Other auto-configuration mechanisms might need to be performed, such as gateway node detection.

If the new neighboring node is a foreign node, it will never be included in the cluster. If a non-capable gateway node discovers a foreign node, no connection will be established. Otherwise, the node is capable of becoming a gateway node and it then adds the foreign node to its list of neighboring foreign nodes. The gateway-capable node can decide to establish a link with the newly discovered foreign node and thereby becomes a gateway node. However, the foreign node will never be able to be included in the cluster. Instead, the gateway node filters out non-authorized traffic from the foreign node, including routing packets and other management packets.

It is also important to detect when neighbors disappear. The link layer may report that a node is no longer connected or otherwise a network layer mechanism based on hello messages can be used to detect nodes disappearing. When a node loses the connectivity to a neighboring personal node, it will remove that node from its list of neighboring personal nodes. However, this does not necessarily mean that the node is no longer a member of the cluster; the node may still be connected to the cluster through a different path. It is then up to the intra-cluster routing protocol to identify a new route. A gateway node will also remove a no longer connected foreign node from its foreign node neighbor list.

5.3.4   Merging and Splitting of Clusters

The inclusion of a single personal node into a cluster is a special case of two clusters merging, since a single isolated personal node forms its own one-node cluster. When two clusters with more than one node merge, the same procedure will take place. The only difference is that the addressing, routing, and auto-configuration protocols may need to do more work, since more than one new personal node is being added to the cluster. The same goes when a cluster splits up into two clusters. Only the addressing, routing and auto-configuration protocols may need to do more work.

5.3.5   Cluster Member List

In this approach, there is no mechanism that keeps track of the member nodes of a cluster. However, other mechanisms, such as the inter-cluster routing protocol, may need this information. In that case, a table-driven intra-cluster protocol can be used. If a node has a route in the routing table to another node, then that node belongs to the same cluster. The routing table will list all nodes in the cluster and this information can be used by the gateway nodes to construct the right inter-cluster tunnels and the correct routing over these tunnels. Another option might be to use similar information from the addressing scheme, if available.

5.4   Intra-Cluster Routing

The cluster formation mechanism finds personal node neighbors and establishes secure connections to them. Hence, it creates intra-cluster links. However, as mentioned earlier, a cluster may become multi-hop, that is, not every personal node in a cluster has a direct link to every other node in the cluster. This means that a routing protocol is required and, due to the dynamics of the topology of a cluster, an ad hoc routing protocol is recommended.

5.4.1   Ad Hoc Routing Protocols

Ad hoc routing protocols are routing protocols specifically designed for ad hoc networks (see Section 3.2). Since a cluster is a form of ad hoc network, we may rely on one of the proposed ad hoc routing protocols, of which there are many. However, we will not discuss or compare ad hoc routing protocols in depth in this book. Below is a list of the best-known and most popular ad hoc routing protocols. For a more complete overview, see Abolhasan et al. (2004).

Destination-Sequenced Distance-Vector Routing (DSDV). DSDV (Perkins and Bhagwat 1994) was one of the first ad hoc routing protocols. It is a proactive distance-vector routing protocol suitable for high mobility. With a new way of using sequence numbers, it can detect and avoid routing loops even when the mobility is very high. The main problem of DSDV is its poor scalability. As the number of nodes and the mobility increase, the overhead quickly becomes too burdensome.

Optimized Link State Routing Protocol (OLSR). In mobile networks, OLSR (Clausen and Jacquet 2003) is a more scalable approach to proactive routing compared to DSDV. OLSR is a link state routing protocol that builds a backbone and only communicates link states belonging to the backbone or links connecting other nodes to the backbone. Most links in the network are ignored and this, in combination with the use of the Multipoint Relay (MPR) optimized flooding protocol to broadcast link states, makes OLSR much more scalable. Still, it provides near optimal routing between all nodes in the network. OLSRv2 (Clausen et al. 2009b) is a newer version of OLSR that uses the common packet format for ad hoc networks (Clausen et al. 2009c), but basically uses the same mechanisms and algorithms.

Dynamic Source Routing (DSR). In DSR (Johnson and Maltz 1996; Johnson et al. 2007), a node should not need to keep any routing information and no route updating information should be distributed if not used. Instead, it is up to the sending node to discover and maintain a route to a destination on demand. The path is explicitly sent together in each data packet, so that all intermediate nodes know where to forward the packet. To discover the path to a destination, a route request packet is flooded through the network using broadcasting. If the destination node receives the route request, it answers by sending a reply packet back via the reverse path. When the sending node receives the reply via the reverse path, it can start to send data along that path by putting each hop of the selected route into the header of the data packets. If a route becomes invalid, a route error packet is sent back to the sender by the intermediate node where the error occurred. The sender then needs to find a new route to the destination by broadcasting a new route request message.

There are a few additional features that can be implemented to try to improve the efficiency of the DSR protocol, such as replying to a route request using cached and overheard routing information, packet salvaging using alternative routes when the selected route fails, and automatic route shortening of existing routes.

Ad Hoc On-Demand Distance Vector (AODV). AODV (Perkins and Royer 1999) is similar to DSR, but the path is not transmitted in the data packets. Instead, the intermediate nodes have to remember the path. This means that each node in the network only must remember all active paths, which is still much less than all possible paths in the network. The route request handling in AODV can be made more efficient than in DSR, since not only the destination node may answer the route request packet. Any node that has an active path to the destination node may answer the route request.

Dynamic MANET On-demand Routing (DYMO). DYMO (Chakeres and Perkins 2009) is the continuation of AODV toward standardization. The main enhancements are a clearer specification and the use of the common packet format for ad hoc networks (Clausen et al. 2009c) mentioned earlier in Section 5.2.

While DSDV and OLSR are proactive routing protocols, DSR, AODV, and DYMO are reactive protocols. The main drawback of reactive protocols is the delay introduced due to the route setup. However, in general, both reactive and proactive routing protocols have their pros and cons. Depending on the network condition (mobility, number of active paths, path durations, etc.), different routing protocols will be best. For good quantitative comparisons between the routing protocols, the reader may wish to look at Haas and Pearlman (2001) or Hoebeke (2007).

5.4.2 Link Quality Assessment

A very large part of the research on ad hoc routing protocols focused on routing signaling and overhead in idealized network environments. However, when such protocols are used in real wireless networks, they often ignore the quality of the links and, as a consequence, choose routing paths with lower quality. To make unicast routing a success in clusters, we need not only an efficient and accurate routing protocol, but also a good approach to correctly assess the quality of the wireless links and thereby be able to identify the paths with the highest quality. Experiments (De Couto et al. 2002, 2003b) have shown that links with packet delivery ratios of 40–80% are common in both indoor and outdoor environments.

In most current implementations, the quality of wireless links is assessed based on the exchange of hello packets. Even though several hello packets have successfully been exchanged, a link may still not be good enough to carry data traffic. The quality of wireless links varies in time to such a degree that it is often not clear whether a link is suitable for data transmission or not; they do not exhibit an on–off behavior. Hence, a better assessment of the link quality is necessary to achieve satisfactory communication within a cluster. Since a cluster may use very different types of links and the devices may be of varying capabilities, there must be such mechanisms for each potential wireless technology as well as a mechanism to compare between the different technologies. We refer to this functionality as *link quality assessment* (LQA).

The goal of LQA is to provide information for maximizing the quality of the end-to-end links within the cluster. For most applications, this means achieving the lowest possible packet loss and delay and maximizing throughput. Hence, the routing protocol must identify the best path for each session while not inflicting too much overhead. This is made quite difficult because typical clusters contain several very different communication technologies and nodes with varying capabilities. At the same time, clusters may be very dynamic with constantly changing conditions.

The LQA mechanism must feed the routing protocol with information about the quality of available links that is as accurate and up to date as possible. The main aim is to predict the quality of a link and preferably also predict the link quality trend. It is the task of the routing protocol to act on these changes. For the LQA function, we need a solution that is able to quickly and accurately detect, or even anticipate, link changes.

Several methods can be considered to assess the quality of the link. All of them include testing the link using hello packets or passively observing the transmission of ongoing data traffic. Below, we discuss the most promising techniques for LQA.

Hello packets. A simple way to predict link quality is to continuously and periodically generate hello packets. If the hello packets are received at a neighboring node, then there is a link. The packet loss can be estimated by transmitting a known number of hello packets over a certain time and observing how many of them arrive, for example, counting how many of the last 10 hello packets have been received. Numerous protocols and implementations use this method. A common procedure is to broadcast a hello packet every X seconds, where X may vary between 1 second and perhaps 10 seconds. By using broadcasting, one transmission can be used to detect the link quality with all neighbors at once and also to discover new neighbors. The link layer usually does not use retransmissions for broadcast traffic and hence the real packet loss ratio before retransmissions can be determined.

In most implementations, the window size for counting the number of hello packets is 10. Supposing that the hello packet interval is 1 second, then it takes 10 seconds for a node to fully learn the quality of the link to a new neighbor. In the same way, it takes this mechanism 10 seconds to fully learn the new quality of a link when it changes. For clusters, which are dynamic in nature, this mechanism reacts too slowly to link changes. If a link suddenly breaks or changes quality, it typically takes several seconds before the routing layer detects this and takes action. When this happens frequently, it causes detectable quality drops and severely degrades the performance of real-time applications running on top of these networks. Hence, the packet delivery ratio measured by hello packets is not accurate and fast enough for clusters. Furthermore, hello packets can be lost due to both bad wireless channel conditions and collisions and this makes it even more difficult to use. It is very difficult to know the exact cause of a packet loss.

Another problem when using hello packets that applies to many link layer technologies, including IEEE 802.11, is the inaccuracy caused by the data rate and packet size differences between hello packets and data packets. Hello packets are typically smaller than data packets and are always sent using broadcasting. Broadcast packets in the IEEE 802.11 protocol family are sent using one of the lowest data rates in order to be backwards compatible (usually 2 Mbps). These two differences will cause the delivery ratio measured by broadcast hello packets to differ from the packet delivery ratio when using the same link for data traffic (Anastasi et al. 2004; Chakeres and Belding-Royer 2002).

Received signal strength. The most commonly used cross-layer information for link quality assessment is either the received signal strength indication (RSSI) or the signal to noise ratio (SNR). In general, the use of signal strength for LQA or as a routing metric has not been very promising. The reason is the weak correlation between signal strength and packet loss. Several studies of this have been done; see, for instance, De Couto et al. (2002) or Aguayo et al. (2004).

Another problem with signal strength is its fluctuating behavior. The wireless channel causes the signal strength to vary even in controlled and stable environments. Furthermore, we only know the signal strength associated with received packets. Hence, the estimation will not be very accurate if there are many packet losses, since we do not know the signal strength associated with the lost packets. Experiments show that it is better to assume that a lost hello packet has the lowest receivable signal strength instead of ignoring it (Zhou et al. 2007b).

Data packet retransmissions. When a link is carrying data traffic, we can gather even more information about its current quality. Hello packets are only sent infrequently, which means that the delivery ratio deduced from hello packets has poor accuracy and reacts slowly to changes. If a link carries a lot of data packets, then the delivery ratio of the data packets can also be used to perform LQA. However, there are three problems with this approach. First, most wireless technologies provide retransmissions for the delivery of data packets and we need the packet delivery ratio including all transmissions and retransmissions. That is, we need the packet delivery ratio before any retransmission scheme. Second, the delivery ratio of the data packets depends on the sizes of the data packets. Large packets have lower chances of successful delivery than small packets, sometimes much lower. Third, if there is no or very little data traffic on a link, then this approach cannot be used, since no data can be collected.

Despite these problems, the data packet delivery ratio provides the best option for LQA, since it measures the actual delivery ratio. It is important that we have good accuracy and fast detection of link quality changes for links that are actually used for data transmission. It is more crucial that the routing protocol can quickly react to a sudden degradation of a link in use in order to maintain a high end-to-end path quality at all times. To quickly react to a quality improvement is usually not required. On the other hand, the chances of the link quality soon degrading again are very high.

Combinations of all. When data is transmitted over a link, we would like to use the feedback from the data packet retransmissions, since it offers the best LQA. However, as fewer data packets are transmitted over a link, more emphasis should be placed on the hello packet delivery ratio and the received signal strength. Research shows that by combining the hello packet delivery ratio and the received signal strength, better LQA can be achieved than using them individually (Zhou et al. 2007b). However, when relying on data packet retransmissions and a combination of hello packets and received signal strength, we need to make sure that they can be compared to each other and that one of them does not always give higher estimates than the other.

It is also very important to point out that a link can have one type of quality in one direction and a completely different quality in the other direction (De Couto et al. 2003a, b). That is, a wireless link can be highly asymmetric. Most routing protocols only use one bidirectional metric for a link instead of one metric for each direction. The main reasons for this are lower overhead and the fact that unicast communications involve packets being sent in both directions anyway. Data packets are sent in the forward direction, while acknowledgments are sent in the reverse direction.

Therefore, it is important to realize what information is available at which end of the link. For signal strength, this is not an issue, since signal strength is known to show good symmetry (the Lorentz reciprocity theorem). Furthermore, signal strength can also be obtained from acknowledgment packets, so even if the data stream is only going in one direction, we can acquire good signal strength information on both sides. However, the packet delivery ratio can be very different in the two directions and this needs to be taken into account.

Furthermore, it is likely that a node supports more than one type of link layer technology for intra-cluster communication. In those cases, the node needs to implement one LQA mechanism for each link layer type. The LQA implementations must be tailored to the characteristics of their respective link layer technology and hence may significantly differ from each other. However, it must be possible to compare LQA results between a particular link layer and the others in a fair way. To achieve this, the LQA results must reflect the behaviors of the link layer techniques.

The different link layer technologies may also have other characteristics that should influence the selection of one path over another. For example, power consumption may be a factor and hence should be reflected in the LQA. However, this makes it necessary to define a tradeoff between power consumption and achievable throughput, but such tradeoffs are difficult to define. Instead, it is probably better to make this tunable or have it decided by a context-aware decision system. Such a system may take battery charge levels and other external factors into account when reconfiguring the routing metrics used.

5.4.3 Unicast Routing

The early ad hoc network protocols, such as AODV (Perkins and Royer 1999) and DSR (Johnson and Maltz 1996), try to find the shortest path between the source and destination and ignore the link quality. They mainly use hello packets to detect neighbors and links. However, this does not lead to the best end-to-end performance. Paths with the minimum number of hops often may include very long links with poor quality, and such links usually create more retransmissions and use a reduced throughput if multi-rate is supported on such links. Research has shown that there is room for improvement (De Couto et al. 2003b). Alternative paths that avoid poor links, even if the paths are longer in terms of number of hops, can achieve better end-to-end performance.

One of the simplest ideas for improvement is to exclude links with bad performance. For instance, links with more than 50% packet loss at the lowest transmission rate and highest transmission power are excluded. Using the remaining links, the routing protocol finds the shortest path as usual. The drawback with this approach is the balance one needs between allowing bad links and partitioning the network. A threshold that is too high breaks the connectivity of the network, and one that is too low allows too many bad links to be used.

A more promising approach is to minimize the number of transmissions required to reach the destination (De Couto et al. 2003a) instead of minimizing the number of hops. This is referred to as expected transmission count (ETX). The ETX of a link is the expected number of retransmissions required to successfully transmit a packet over it. Hence, if a link experiences a packet delivery ratio of p, then its ETX is $1/p$ (for a link with 25% packet delivery ratio, we expect to need to resend $1/0.25 = 4$ times, hence its ETX is 4). The routing protocol then needs to find the path with the lowest ETX, where the ETX of a path is the sum of the ETXs of its links. This will lead to the one with the least amount of transmissions, including retransmissions, being used to deliver each packet to its destination; this will increase throughput and minimize delay. ETX will find paths with a good balance between the number of hops and the number of retransmissions per hop.

Another promising alternative is to minimize the time spent occupying the shared medium with transmissions to reach the destination, which is known as expected transmission time (ETT) (Awerbuch et al. 2003; Bicket et al. 2005). In Awerbuch et al. (2003) the metric is actually referred to as the medium time metric (MTM), but the two proposals are almost identical. Furthermore, ETT is similar to ETX except that it also considers the link rate. If all links use the same rate, ETX and ETT are in fact the same. Hence, in networks with multiple link-layer technologies using different transmission rates or when using multi-rate link-layer technologies, ETT is better. ETT can find a good balance between link rates, retransmissions, and number of hops.

Both ETX and ETT find paths with expected minimum required transmissions including expected retransmissions or minimum required 'channel time'. Both will work well in moderate sized clusters, since most nodes are within contention range. Only very large ad hoc networks can benefit from approaches where one tries to route around areas with a lot of contention (Heijenk and Liu 2006). However, if different links can use different non-overlapping channels, it might be advantageous to use weighted cumulative ETT (WCETT) (Draves et al. 2004). WCETT also accounts for the interference between the links. It chooses as much as possible links that use channels that do not interfere with each other in order to maximize the end-to-end throughput.

ETX, ETT, and WCETT all need good LQA information about the links in the network, and, as mentioned in Section 5.4.2, simple hello packets are usually not enough. The routing mechanism needs to acquire this information from the links. Hence, for good intra-cluster routing, we need routing that uses good quality LQA as input. Research into this for small ad hoc networks, such as typical clusters, has been reported in Zhou et al. (2007a, 2008, 2009).

5.4.4   Cluster-Wide Broadcasting

PNs need cluster-wide broadcasting. For many applications, it is necessary to transmit a packet from one node to all the others inside a cluster. However, in a multi-hop cluster, the range of the source node does not necessarily cover the whole cluster and the node therefore needs help from other nodes to relay the packet to reach all nodes. This process is called flooding and is an important feature for any wireless multi-hop network.

Flooding is used by several unicast routing protocols (Johnson et al. 2007; Perkins et al. 2003) for multi-hop networks to disseminate route requests or link states. Other applications include service discovery, sharing of context information, address autoconfiguration (CempakaWangi et al. 2008), and network self-organization. In mobile networks, flooding is often a better choice than multicasting due to the frequent topology updates (Obraczka et al. 2001).

The simplest flooding mechanism for multi-hop networks is blind flooding, in which a node always retransmits a received flooding packet after a small random delay (jitter). Each node needs to keep a list of recently received flooding packets to detect duplicates and avoid retransmitting the same packet twice. However, this is the only mechanism in blind flooding that reduces the number of retransmissions.

Several studies of flooding protocols have revealed that it is possible to reduce the number of retransmissions much more (Tseng et al. 2002; Williams and Camp 2002). This has the benefit of reducing contention and collisions, saving energy, and it may even increase the flooding speed. Such optimized flooding protocols can make use of location information, neighbor information, or observations of the ongoing flooding. Protocols using location information require solutions such as Global Positioning System (GPS) in every node, which is too strong a requirement for most PN devices. Besides, the improvements brought by location information have not proved to be very significant (Williams and Camp 2002), especially if there is mobility. Hence, solutions that do not use such information are preferred.

Many different flooding protocols have been proposed in the past. For an extensive survey of flooding protocols, see Williams and Camp (2002) or Stojmenović and Wu (2004). The IETF MANET working group has also started a standardization effort targeted for flooding in multi-hop networks (Macker 2009). This work is just a framework for flooding, leaving the actual choice of flooding protocol open. That is, the framework only specifies mechanisms related to packet formats, neighborhood information collection, and how nodes should detect that a received packet is a duplicate. Several examples of flooding protocols are given in the document together with how they could be implemented using the framework.

5.5 Summary

In this chapter, we have defined a cluster as a connected network consisting exclusively of active personal nodes and shown that this is the best choice. It makes intra-PN communication rely on local communication to the largest possible degree. Since we believe local communication will outperform long distance communication most of the time, this should lead to the optimal solution.

Clusters are nothing more than a set of active personal nodes that can communicate with each other using their own communication features and without external support. Clusters are basically ad hoc networks and may therefore consist of multiple heterogeneous wireless technologies and hence be multi-hop networks.

We introduced mobile ad hoc network technology and discussed how it can be used for clusters. Finally, we walked through all the necessary mechanisms for cluster formation and maintenance, such as discovery of neighboring personal nodes, link quality assessment, intra-cluster unicast routing, and cluster-wide broadcasting.

6

Inter-Cluster Tunneling and Routing

To achieve communication between all of a person's personal devices, intra-cluster communication is not enough. We also need to solve the issue of communication between personal nodes in different clusters, which we call inter-cluster communication, and that is the topic of this chapter.

As we explained in the previous chapters, a PN typically has multiple clusters that are geographically dispersed. To support inter-cluster communication, the clusters must first have access to a fixed interconnecting structure through one or multiple gateway nodes. To realize communication between the clusters and form a PN, clusters need to locate each other, establish connections and routes between each other, and maintain all this regardless of cluster or node mobility (Jacobsson et al. 2004).

The discovery and selection of gateway nodes that can provide access to the fixed infrastructure is a cluster-internal issue and is best fulfilled through a cluster-internal mechanism. For the clusters to be able to locate each other, we introduce the concept of a PN agent. Clusters that have obtained access to the interconnecting structure announce their presence to the PN agent by sending a packet as shown in Figure 6.1(a). The announcement packets contain information, such as security credentials, current care-of addresses (CoAs) for the gateway nodes, and perhaps also the list of nodes in the cluster. The PN agent can communicate this information to other clusters and their gateway nodes, which may trigger the creation of secure tunnels between the clusters as shown in Figure 6.1(b).

The purpose of the tunnels is twofold. First, they provide secure inter-cluster communication by shielding the intra-PN communication from the outside world. Second, these tunnels are dynamically updated in order to handle the mobility of clusters.

To handle the mobility, information regarding the change of availability of gateway nodes is propagated and the gateway nodes will react to this by dynamically updating the tunnels. When a gateway node changes its attachment point to an interconnecting structure, existing tunnels are terminated and new ones are created. In cooperation with PN-wide routing and addressing, this results in a self-organized PN that consists of several clusters interconnected by dynamic tunnels. This provides security and hides the cluster mobility and gateway node changes from the nodes in the PN.

Personal Networks: Wireless Networking for Personal Devices   Martin Jacobsson, Ignas Niemegeers and Sonia Heemstra de Groot
© 2010 John Wiley & Sons, Ltd

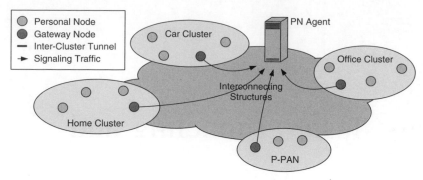

(a) Gateway nodes inform the PN agent.

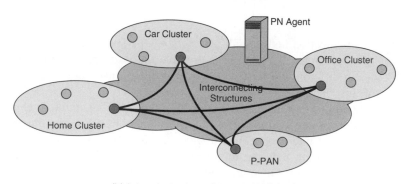

(b) Inter-cluster tunnels are established.

Figure 6.1   Example of PN establishment.

When Jane is in her office, she can access her files using her desktop computer. However, she does not need to copy those files to her laptop before leaving for a company visit. Instead, her laptop can connect to her office cluster and her desktop computer using the UMTS link in her mobile phone. Her mobile phone acts as a gateway node and establishes a tunnel to the desktop computer in her office, which acts as a gateway node for her office cluster. As long as she carries her phone, she can access any file in her PN. In fact, there is smart and distributed file system software active in her PN that automatically distributes her files in such a way that they are always accessible. Furthermore, backups are kept, reducing the risk of loosing valuable data.

When Jane arrives at a client site, her cluster automatically reconfigures its connections to the office cluster. The client has allowed her to use their Internet connection, so her cluster deactivates the UMTS connection in her phone and connects the laptop directly to a WLAN access point hosted by the client. Hence, the PN system automatically configures the laptop to become the active gateway node instead. This saves battery power in her phone and at the same time gives Jane higher throughput. With this new link, she can even stream multimedia content from her office to a projector at the client's site for display to the people there.

This chapter is organized as follows. Section 6.1 introduces important requirements for inter-cluster communication. Section 6.2 discusses general IP mobility solutions and how they apply to PNs. In Section 6.3, we cover PN-wide addressing. Section 6.4 discusses aspects of infrastructure support. Section 6.5 covers the main functionalities of inter-cluster tunneling, including mobility handling, tunneling strategies, gateway node coordination, and security. In Section 6.6, we continue with aspects of PN routing, including PN-wide broadcasting and QoS. We conclude with a summary in Section 6.7.

6.1 Inter-Cluster Tunneling Requirements

For a successful implementation of inter-cluster communication, it is crucial that the following requirements are met:

 (i) Once access to the interconnecting structure is available, the clusters need to be able to communicate and thus form a complete PN. Routing between all the nodes in all the clusters must be supported.

 (ii) As much of the communication setup as possible must be automatic. It is unacceptable to require the user to manually bring connections up and down. Everything related to inter-cluster communication must be done automatically and intelligently by the system. However, some user control should be possible, such as permitting the use of access networks that are charged. The user may specify rules by which decisions are automatically made by the system.

 (iii) The inter-cluster communication must support mobility. This includes handovers between access networks as well as handovers between gateway nodes, whenever better access networks arise or those in use disappear.

 (iv) As many existing access networks and interconnecting structures as possible must be supported. Emerging and future technologies should be supported, inasmuch as they can be anticipated. This means that current deployment setups must be supported, such as WLAN hotspots and digital subscriber lines (DSL) using IPv4, Dynamic Host Configuration Protocol (DHCP) (Droms 1997), and network address translators (NATs). However, as future deployments evolve (e.g. using IPv6), they must also be supported. Hence, a great deal of flexibility is required.

 (v) It must not be necessary to change or add functionality to the current interconnecting structures. The deployment of PNs must not depend on certain technologies first being deployed in the infrastructure. This is not to say that additional functionalities that can improve the operation of PNs are not wanted, just that such functionalities must not be required, since that would hamper the adoption of PNs.

 (vi) The user's privacy must be retained. This is important, since a PN will carry user-related data which will be transmitted over various access networks and interconnecting structures. Hence, all inter-cluster communication must be properly protected with both data confidentiality and data integrity.

 (vii) It is also important that inter-cluster communication can achieve a high QoS whenever required. This implies that the PN must be able to choose the best gateway node, access network type, and access point for communication. The quality of every option should, to some reasonable degree, be monitored or predicted so that a good selection can be made. The selection may also be based on the needs of the applications active at any moment.

(viii) The inter-cluster communication mechanisms should not waste energy or other resources. Unnecessary communication paths that consume energy or resources should be disconnected when not needed. This requirement may be in contradiction with the previous requirement on QoS, so a proper tradeoff must be made.

In the remaining sections of this chapter, we will introduce and discuss options and solutions that can successfully meet all these requirements.

6.2   IP Mobility

Many proposals for mobility support for IP have been proposed. In this section, we will investigate them and see how they can be applied to inter-cluster communication.

6.2.1   IETF Network Layer-Based Proposals

The most important mobility solutions for IP networks are IP Mobility Support for IPv4 (Perkins 2002) and IP Mobility Support for IPv6 (Johnson et al. 2004). Both Mobile IPv4 and Mobile IPv6 require every mobile host to have a so-called home agent (HA) on their home network. Any host that wants to communicate with the mobile host sends packets to the home address, which is the address of the mobile host when it is on the home network. The HA intercepts those messages and forwards them in a tunnel to the mobile host, either directly or via a foreign agent (FA). The mobile host updates the HA when its care-of address (CoA) changes, by sending a message with the new address. However, packets sent by the mobile host in the other direction are sent directly and not via the HA. This results in triangular routing, involving the corresponding host, the HA, and the mobile host. However, in Mobile IPv6, it is also possible to send the CoA to the corresponding host so that all traffic can flow directly between the two hosts, thus avoiding triangular routing.

Mobile IP is concerned with mobility of single hosts, such as laptops. Hence, each node has its own mobility mechanism. If applied to a cluster, which is a network of mobile hosts, each node must have its own mobility mechanism. When a cluster roams, all nodes need to update their HAs concurrently when the network changes its point of attachment. For PNs, this will lead to a lot of signaling when a cluster changes it attachment point to the interconnecting structure. Furthermore, a PN does not have a home network where its HA can be located.

The Network Mobility (NEMO) Basic Support Protocol (Devarapalli et al. 2005) was introduced to address the issue of network mobility. In NEMO, each mobile network has a mobile router (MR). The MR is connected to the infrastructure and has a home agent just as in Mobile IPv6 (NEMO is only specified for IPv6). Instead of a home address coupled to the home location, the MR has a home network prefix coupled to the home network. All the nodes in the mobile network get addresses with that network prefix from the MR via link local IPv6 router advertisements (Narten et al. 2007). The nodes in the mobile network do not need to be aware of their mobility when the network roams, only the MR needs to take actions. In NEMO basic support, all traffic to and from the mobile network is tunneled via the HA. However, route optimization has been proposed (Ng et al. 2007b).

NEMO has several limitations and drawbacks for PNs. First, the mobile network needs to be a single-hop network since it relies on link local router advertisements. Hence, NEMO cannot support multi-hop clusters without additional functionality. The only option would be to use Ananas (Chelius and Fleury 2002), which makes a multi-hop network behave like a normal single-hop LAN for the IP layer by introducing an intermediate layer. As described in Section 5.2, Ananas is not ideal and hence a better solution is to replace the network prefix advertisements with a solution that works over a multi-hop network. Until very recently, neither NEMO nor Mobile IPv6 supported multi-homing. However, the IETF has now made a proposal for this (Ng et al. 2007a; Wakikawa et al. 2009), thus enabling soft handovers between two access networks as well as multi-homing for increased reliability and load sharing. However, in NEMO, there is the additional problem of multiple MRs in a single mobile network. If the two MRs have different network prefixes, then it becomes difficult for mobile nodes to switch from one MR to the other. Hence, when using NEMO for PNs, all the MRs belonging to the PN need to have the same prefix. Unfortunately, this is difficult to achieve as the HA would not know which MR to forward a packet to when the MRs are no longer in the same mobile network (cluster). Because of these limitations in NEMO, we do not propose to use NEMO for inter-cluster communication.

The IETF's work on mobile nodes and networks is currently quite intense. Another important area that it is focusing on is that of detecting network access (DNA) (Choi and Daley 2005; Krishnan and Daley 2009). Whenever a mobile node discovers access to a new network on one of its interfaces, it needs to detect the IP configuration of that network as well. Unfortunately, the IPv6 neighbor discovery protocol (or the DHCP for IPv4) is not very fast and this is what DNA is trying to remedy. It is also worth mentioning that timely support from the link layer, similar to our work in Section 5.4.2, is crucial for achieving smooth handovers. We will discuss this topic in Section 6.6.2. Nevertheless, as soon as DNA becomes available, gateway nodes should start using it when connecting to the interconnecting structures.

Another potential candidate for inter-cluster communication is the Host Identity Protocol (HIP) (Moskowitz et al. 2008). In HIP, each host has at least one fixed identifier called a host identifier (HI). The HI may have the same form as an IP address and is used when establishing TCP or UDP sessions. The HI is mapped to a current valid and topologically correct IP address of the host. To initiate a new HIP session, a corresponding host would need to know the host's HI and its current IP address. This can be discovered if they are direct neighbors on the same network. If not, the host may have a rendezvous server (RVS) (Laganier and Egger 2008), which is published together with the HI in, for instance, DNS. Each host will keep its RVS up to date by informing it through HIP messages every time it changes IP address (Laganier et al. 2008). When an HIP session is established with another host, a change of the IP address is also communicated to the other host by an HIP message (Nikander et al. 2008). This will speed up the mobility handling for the ongoing session. Extensions for NAT traversal also exist (Komu et al. 2009).

The data traffic of HIP is typically tunneled using Internet Protocol Security (IPsec) (Kent and Seo 2005) with Encapsulating Security Payload (ESP) (Jokela et al. 2008; Kent 2005), but other alternatives are possible. Furthermore, this is not the only security feature of HIP. For instance, the HI is actually the public part of an asymmetric key. Using this feature, the end hosts can be authenticated and all HIP messages and data traffic can be protected after a four-way handshake between the two hosts. Unfortunately for PNs, HIP has its own security mechanism, which significantly differs from that proposed for PNs. Hence, for use in PNs, the HIP security mechanisms must be extended or replaced to be used in a PN. Furthermore, HIP only does host-to-host communication and not network-to-network communication. Hence, for PN inter-cluster communication, HIP must be extended with routing over the HIP sessions, which need to become HIP tunnels. Nevertheless, many of the HIP mechanisms can be utilized in one way or another for inter-cluster communication, such as mobility handling, the RVS mechanisms, and NAT traversal.

6.2.2   Other Network Layer-Based Proposals

MOPED (Kravets et al. 2001), which was introduced in Section 3.10, provides a more complete alternative for inter-cluster communication. In MOPED, a person's personal devices, which can connect directly with each other, form so-called components. Components are essentially the same as clusters. A proxy server located somewhere in the infrastructure keeps track of all the personal devices, components, and how they are connected. The proxy knows where each device is and can therefore solve issues such as addressing, routing, load balancing, and mobility. The mobility between the proxy and the perimeter (gateway node in PN terms) of the components is done through low overhead tunnels whose mobility is handled by Mobile IP. A MOPED forms a star topology with the components at the edges and the proxy in the center, which means that all inter-component traffic goes through the proxy. Route optimizations are discussed, but only for traffic going in or out of a MOPED. Furthermore, no security or privacy solutions are discussed. Otherwise, MOPED provides most of the functionalities required for inter-cluster communication in PNs.

Peer-to-peer techniques, such as Chord (Stoica et al. 2003) and Pastry (Rowstron and Druschel 2001), may also provide a good alternative for inter-cluster communication. One solution built on peer-to-peer techniques is Robust Overlay Architecture for Mobility (ROAM) (Zhuang et al. 2003), which is based on Internet Indirection Infrastructure (i3) (Stoica et al. 2002). In ROAM, servers in the infrastructure relay packets to the correct destination using special tags. Each host has a unique tag that is associated with one of the servers. Mobility is supported when the hosts update their destination addresses associated with their tags at the ROAM/i3 servers. ROAM can be used for handling inter-cluster mobility, but the non-direct routing via the ROAM/i3 servers still remains a question.

6.2.3   Application Layer-Based Mobility Proposals

Mobility can also be handled at the transport layer (Ansari and Sathyanath 2007; Maltz and Bhagwat 1998; Snoeren and Balakrishnan 2000; Zandy and Miller 2002) or the application layer (Schulzrinne and Wedlund 2000). The principles are the same as for the various IP-level mobility solutions. To maintain ongoing sessions, either a proxy is inserted between the two peers as in MSOCKS (Maltz and Bhagwat 1998), or redirection messages are sent between the peers as in Migrate (Snoeren and Balakrishnan 2000), STEM (Ansari and Sathyanath 2007), and Reliable Sockets/Packets (Zandy and Miller 2002). The main difference lies in the implementation, whether both TCP and UDP are supported, and whether modifications to the applications and/or the operating system are needed. To locate a mobile node and discover its current IP address, most protocols suggest using the DNS UPDATE protocol (Vixie et al. 1997). However, when using SIP (Schulzrinne and Wedlund 2000), it can be used for both maintaining ongoing sessions during mobility and discovering the current location of mobile nodes. As with the network layer mobility support protocols, any of these protocols could be a part of the solution for the inter-cluster communication framework. However, none of them is well suited for PNs, since they all focus on host-to-host communication, rather than network-to-network.

6.3   PN Addressing

Before tackling PN-wide inter-cluster communication, it is necessary to have internal addressing of personal nodes. Each node should have a fixed intra-PN address that stays the same as long as the node is part of the PN. Since we assume that personal nodes can roam freely, there is no possibility for a hierarchical organization of intra-PN addresses without introducing address changes. If the intra-PN addresses still change, then the mobility problem is not entirely solved. Hence, the intra-PN address should remain fixed.

The only remaining problem is the assignment of these addresses, which must happen as soon as a node joins the PN. The address can be self-assigned, but must be unique within the PN. Worldwide unique identities or addresses, such as the EUI-64 (IEEE 1997), can be used as intra-PN addresses if they are not too long and if the node has one. However, we do not really want to use such globally unique addresses when there is no need for them. Furthermore, 64-bit (or 48-bit) node addresses are unnecessary long for our purpose, if we assume that PNs might have no more than hundreds of nodes. An 8- or 16-bit node address plus prefix is more suitable. A longer prefix will allow for better compression in routing and neighbor discovery packets based on the common MANET signaling packet format (Clausen et al. 2009c).

If we want a short node address, the chance of address conflicts becomes significant. Hence, there is a need for duplicate address detection (DAD). The first step a new node should take, after becoming a member of the PN and assigning itself an intra-PN address, is to verify the uniqueness of its new intra-PN address. Depending on how the personalization is done, it may be possible to use it to also assign unique addresses to the nodes. If a central personalization device is used (see Section 9.1), it may have a pool of non-assigned addresses which can be used and with guaranteed uniqueness within the PN. Another option is to use the PN agent. The PN agent should know about every node in the PN. If a personal node is unknown to the PN agent, it is for practical purposes not yet part of the PN. A last resort is to rely on one of the DAD schemes for MANETs (CempakaWangi et al. 2008). However, they all are much more complex and create much more overhead compared to just querying the personalization device or the PN agent.

6.4   Infrastructure Support

We make use of interconnecting structures to connect the clusters. The only question is how much more support is required compared to what is currently offered (Prasad et al. 2005). IP transport between gateway nodes that are connected to the interconnecting structures is essential, but support from special servers may also benefit the operation of a PN. A contactable server somewhere in the interconnecting structures can offer services similar to a home agent in Mobile IP or a rendezvous server in HIP. We refer to such servers as *PN agents*. Furthermore, the burden of gateway nodes may be reduced by special functionality offered by the access router that the gateway node connects to. Such routers, with special PN functionality, we refer to as edge routers, because they are situated on the edge of an interconnecting structure. In the following, we will discuss the potential roles of the PN agent and the edge routers.

6.4.1 PN Agent

The PN agent is a management entity located anywhere in the interconnecting structure or elsewhere from where it can be reached at all times. We may think of servers permanently connected to interconnecting structures or even a non-mobile gateway node. When relying on PN agents, each PN has its own agent and its task is to keep track of all personal nodes and clusters in a PN. All the gateway nodes need to be aware of the IP address of their PN agent. Therefore, the address is distributed to every personal node during the personalization. Gateway nodes know the address of the PN agent, since they also are personal nodes.

Clusters that have obtained access to the interconnecting structure announce their presence to the PN agent as shown in Figure 6.1. More precisely, the gateway node sends a registration message to the PN agent. The information contained in the registration messages must be transferred in a secure way so that the information in the messages cannot be altered and is invisible to non-authorized parties. The registration messages need to contain at least the following essential information: PN identification, node identification (e.g. intra-PN address), and the CoAs of all the gateway node's active attachment points to the interconnecting structure. The PN identification is needed since there might be more than one PN agent running on one server. The PN agent must also be able to check the credentials of the gateway node to access a certain PN and the message's authenticity. The gateway node's CoAs are of course needed, as they will represent the endpoints of the inter-cluster tunnels. The PN agent stores this information from all the gateway nodes in a secure database.

The information stored in the PN agent can be queried by the gateway nodes in order to establish the inter-cluster tunnels. Alternatively, the PN agent may decide which tunnels should be established or maintained. The PN agent can also assist in establishing tunnels between two gateway nodes that cannot directly establish a tunnel due to firewalls or NATs and, if even that is impossible, the two gateway nodes may send their data traffic via the PN agent.

The purpose of the PN agent is quite similar to that of the home agent in Mobile IP or the rendezvous server in HIP Host Identity Protocol (HIP). In this light, the PN agent is best seen as an abstract entity and can be based on either Mobile IP or HIP. Mobile IP might not be the best choice, due to the difficulty of achieving direct tunnels between gateway nodes without modifying the protocol. HIP, on the other hand, implements functionalities that we do not need. In the following section, we will sketch a protocol based on a simplified version of HIP, where the security parts of HIP are replaced with the PN security based on personalization.

Note that the PN agent also can provide functionality other than networking. It may assist in other PN-internal mechanisms, such as name resolution and service discovery. Further, the PN agent can be used by foreign nodes that wish to communicate with the PN. In that case, the address of the PN agent is the only address a foreign node needs to know in order to be able to communicate with the PN. This will be discussed further in Chapter 7, where we will address the communication between PNs and foreign nodes.

Even though we have described the PN agent as running on a single node in the interconnecting structure, it is not necessarily so. The PN agent could be a distributed functionality running on several servers, a set of redundant servers (Tuexen et al. 2002), or a peer-to-peer network of servers, such as Chord (Stoica et al. 2003). There can be many reasons for a distributed PN agent, but the two most important reasons are increasing availability of the PN agent functionality and reducing response time experienced by the gateway nodes.

Another important aspect is where the PN agent server or servers should reside and under whose control and responsibility they are. A user who needs or wants total control of the PN agent may wish to run the PN agent functionality on one of his own nodes, such as a gateway node in the home cluster that connects to the interconnecting structures using a reliable and fixed connection without NAT. Alternatively, network or service providers may offer PN agent functionalities that can be used by their customers. Hence, there are several options for PN agent deployment, and which option is best not only is a technical matter, but also has, for instance, business consequences.

6.4.2   Edge Routers

An edge router (ER) is an access router that sits on the edge of the interconnecting structure, communicates with the gateway nodes, and supports them by offering special PN functionality (Hoebeke et al. 2006b; Louati and Zeghlache 2005; MAGNET 2004a). They need to be managed by the network access provider and thus will probably be owned by the provider. On behalf of a cluster, an ER can perform several intra-PN tasks, such as communicating with the PN agent and taking care of the inter-cluster tunnel establishment and management. In this way, ERs can relieve the gateway nodes of some of the work and thus allow them to reduce their power consumption and resource requirements. Figure 6.2 shows an example of inter-cluster communication using ERs. In this example, we assume that not all access networks provide ER functionality and hence some gateway nodes still need to perform all tasks related to inter-cluster tunneling (the gateway node in the car cluster).

If we assume that gateway nodes need to maintain many tunnels, then this maintenance consumes valuable resources, such as processing and energy. The tunnel maintenance may therefore overload the gateway nodes, which are often mobile and battery-powered. Thus, it will be useful to let ERs, which are fixed and powerful, support the tunnel establishment as much as possible and thereby place the overhead needed for establishing and operating a PN in the interconnecting structures. Furthermore, ERs can assume other responsibilities – inter-cluster routing, remote service discovery, service repository, and more.

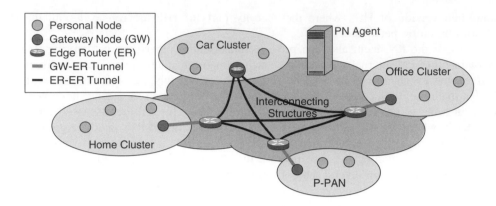

Figure 6.2 Inter-cluster communication with edge routers.

The use of ERs has both advantages and disadvantages (Prasad et al. 2005). The advantages of an ER-based solution can be summarized as follows:

 (i) Some tasks can be carried out by the ER, which leads to less consumption of scarce resources in mobile devices.
 (ii) Mobile gateway nodes can be made more lightweight. This leads to simpler devices that have lower cost and less power consumption. Consequently, more PN nodes can provide gateway node functionality, which leads to increased flexibility in accessing the interconnecting structures.
(iii) PN formation and maintenance are faster and hence can better support cluster mobility.
(iv) ERs may support special functionality that can optimize handovers that current access technologies do not offer. This could, for instance, include Fast Handover for Mobile IPv6 (FMIPv6) (Koodli 2009) and/or Hierarchical Mobile IPv6 Mobility Management (HMIPv6) (Soliman et al. 2008).

The use of ERs has the following drawbacks:

 (i) ERs need to be deployed. Since ERs are access routers specifically designed for PNs, it is necessary to modify the infrastructure by introducing these network elements. Tasks like this have proven to be difficult in the past, and that is likely to be a major stumbling block. If gateway nodes require ERs, then the deployment of PNs can only take place after network providers have invested on a sufficient scale in ERs and hence there is a risk that this will slow down the success of PNs. Furthermore, the network providers need to maintain these more complex ERs.
 (ii) ERs do not reduce the complexity of the PNs. Due to the expectation that there will be many access networks without ERs, it is still necessary for the gateway nodes to implement full gateway node functionality. As long as not all access networks offer ER functionality, this will be a drawback. In the meantime, gateway nodes need to handle two cases: access networks with ERs and access networks without ERs.

(iii) The ERs need to be trusted by the user, since ERs will support the internal mechanisms of the PN, and this may endanger the security and privacy of the PN. If an ER is not trusted, the gateway node cannot use the ER and must instead perform all gateway functionalities itself. Furthermore, there must be trust between the ERs and the PN agents, even when they belong to another network provider or to the PN user.

The issue of ER deployment is a major drawback that currently makes it important to support ER-less access networks. Future solutions may consider ER technologies as a way to optimize the performance of PNs. However, for the time being, ER-less solutions are more important.

6.4.3   PN Networking without Infrastructure Support

It is also possible to design an intra-PN communication system without special infrastructure support such as ERs and PN agents. To do so, we need to turn our attention to peer-to-peer technology. The biggest advantage of peer-to-peer technology is that, in principle, it can make infrastructure-based support unnecessary. The ROAM (Zhuang et al. 2003) peer-to-peer system, which we introduced in Section 6.2, unfortunately does not demonstrate this advantage. Instead, ROAM requires the deployment of i3 servers and it is these servers that form the peer-to-peer network.

To make both ERs and PN agents superfluous, we need to make the gateway nodes themselves into peers in the peer-to-peer network. This means that the gateway nodes need to manage the inter-cluster tunnels themselves. Each gateway node needs to remember the current locations (i.e. the CoA) of as many other gateway nodes in the PN as possible. This will work if some gateway nodes almost never move or if not all gateway nodes move at the same time. Normally, we can expect that at least one gateway node (e.g. the home cluster gateway node) never moves, which means that the other gateway nodes always can connect to that gateway node to be updated.

There are two main problems with not having the support of a PN agent. The first major problem is the bootstrapping of the peer-to-peer overlay. In the beginning, all personal nodes need to gather in one single place and form one single cluster in order to exchange the CoAs. New gateway-capable nodes added to the PN need to communicate with a connected and updated gateway node in order to retrieve all current CoAs. Whenever a node has been deactivated for a long time, this procedure might need to be repeated. There is always the risk of some of the gateway nodes being disconnected from the rest and not knowing how to reconnect to the PN, even though it may work most of the time.

The second major problem in PNs without a PN agent is the slow response to mobility. When a tunnel needs to be updated because an endpoint has moved, it is important that a new CoA or an alternative gateway node is found quickly. Unfortunately, when several other gateway nodes have also disappeared, it may take a while before the right gateway node is queried. The alternative is to update all the other gateway nodes all the time, which, of course, introduces a lot of overhead. For these reasons, we propose to always make use of a PN agent. In Section 7.2.2, we will discuss yet another good reason for PN agents, namely that they provide an easy solution for foreign nodes wishing to establish a connection with the PN.

6.5 Inter-Cluster Tunneling

Inter-cluster tunneling can be done in several ways. The main requirement is that intra-PN packets are encapsulated in encrypted IP packets that travel over the interconnecting structures between gateway nodes. In addition to this, the packet overhead should be minimal and the encryption lightweight. A good option is to use IPsec with UDP-encapsulated ESP (Huttunen et al. 2005) or a similar approach. The keys should be derived from the personalization step. See Chapter 9 for more details on this.

Whenever a gateway node discovers an attachment point to an interconnecting structure, it creates a *tunnel endpoint* (TEP) for that attachment point. A tunnel is a connection between two TEPs that belongs to two different gateway nodes in two different clusters. A TEP is nothing more than a CoA belonging to an active interface on a gateway node. The CoA needs to be topologically correct for its attachment point to the interconnecting structure so that packets sent to its address find their way to the gateway node. Extra parameters can be associated with a TEP, such as NAT information and expected QoS. The TEP information is shared among the gateway nodes and the PN agent so that tunnels can be established. In this section, we will discuss questions related to which tunnels to establish between which TEPs, how to communicate TEP updates efficiently, how to handle handovers and mobility, as well as NATs and security.

6.5.1 Mobility and Dynamic Tunneling

Every gateway node informs the PN agent about all its active TEPs. If a gateway node has more than one valid IP address for a given active interface, it may announce all of them as TEPs, if it makes sense. Reasons for this may be that the different IP addresses use different paths through the interconnecting structures or that an access network supports both IPv4 and IPv6. Consequently, it is normal for a gateway node to announce several TEPs, especially if the node is multi-homed. However, each TEP must be routable through the interconnecting structures so that it can be used by the other clusters.

A PN agent should keep a complete database of all active gateway nodes in the PN and their active TEPs. Any gateway node may query this database for information about the other clusters, gateway nodes, and their valid TEPs. They should also be able to subscribe to updates regarding other clusters and nodes so that they can receive updates in a timely fashion. In addition to the complete database kept at the PN agent, each gateway node should keep a partial database of the TEPs and clusters related to all its established tunnels. Figure 6.3 shows an example of how the TEP information should be distributed over the PN agent and the gateway nodes. In this example, the TEPs consist of only the CoA (e.g. A11 or A32). In addition to the TEP information, each gateway node may also keep information related to the tunnels, including security keys, QoS, and not yet delivered packets.

The gateway nodes use the TEP information to establish tunnels among themselves. TEP information is exchanged periodically and when changes occur between the gateway nodes and the PN agent. Since connections may suddenly disappear, all TEPs are soft states and need to be updated periodically. However, when possible, changes to the active TEPs must be announced in a timely fashion to concerned parties within the PN. A gateway node that sees one of its TEPs disappearing or a new TEP appearing must

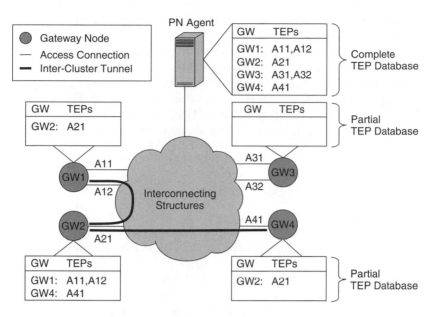

Figure 6.3 Inter-cluster tunneling and TEP databases.

always inform the PN agent about the change. Other gateway nodes in the PN may automatically receive information through the PN agent. To speed things up further and to improve reliability, the gateway node may send the update information directly to the gateway nodes with which it has established tunnels. In this way, fewer data packets will be affected due to the speedy routing of signaling packets. Figure 6.4 shows an example

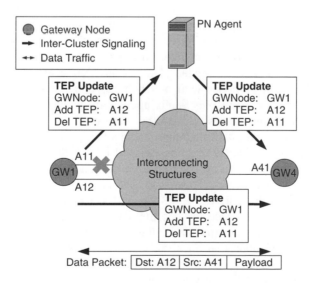

Figure 6.4 Inter-cluster tunnel update due to mobility.

of how the updates are communicated when a gateway node (GW1) looses one of its TEPs (A11) and finds a substitute (A12).

When a gateway node receives TEP updates from other gateway nodes or from the PN agent, it may need to take certain actions. It may need to disconnect an active tunnel and establish an alternative tunnel to an alternative TEP on the same gateway node or to another TEP on another gateway node in the same cluster. If the TEP update contains new TEPs, it may want to switch to the new TEP, for example, because the expected QoS is better. All this is up to the tunneling strategy, which may be influenced by the user's preferences.

6.5.2   Always-Up and On-Demand Tunneling

In an always-up tunneling approach, all possible inter-cluster tunnels will be up and running at all times, even if there is no traffic. This is a proactive approach and requires all gateway nodes to have a complete copy of the TEP database with all the TEPs of all the gateway nodes.

In this strategy, there can be several tunnels between two clusters or even between two gateway nodes. Multiple tunnels may be useful for redundancy, fault tolerance, and increased throughput. These arguments, in combination with the simplicity of the approach, favor an always-up tunnel maintenance policy in which tunnels are established and maintained as soon as a cluster is connected to the interconnecting structure.

The gateway nodes initiate the tunnels with the help of the PN agent, build a quasi-permanent connection with all present gateway nodes in the PN, and keep these tunnels intact as long as possible. When the attachment point of a gateway node changes due to mobility or other reasons, it causes all the tunnels using the old TEP to be diverted to the new TEP. The idea here is to maintain the tunnels proactively between all gateway nodes so that there are valid tunnels between all clusters at all times. The routing scheme determines which tunnels actually will be used and may be based on the QoS information.

The alternative to always-up tunneling is on-demand tunneling, which is a reactive approach. Figure 6.5 shows the difference between these two types of tunneling approaches. In the case of on-demand inter-cluster tunneling, tunnels between gateway nodes are only established when needed. This means that a gateway node only sends TEP updates to the gateway nodes with which it has active tunnels in addition to the PN agent. Further, the PN agent only sends updates to gateway nodes that need them. Except for this, the two approaches work in the same way.

The difference between an established tunnel and a pair of TEPs without an established tunnel is actually not that great. With an established tunnel, the two gateway nodes need to make sure that their session keys (the security associations when using IPsec-based tunneling) are installed and being rekeyed and that TEP updates are communicated correctly and in a timely fashion between the two nodes. However, the greatest difference is whether the gateway nodes need to keep the access network of the TEP up and running. There might be significant power savings if access network connections can be disconnected. A gateway node that does not have any established tunnels may be able to disconnect all its access connections. However, one must not allow all gateway nodes in a cluster to disconnect all their access networks, since this will put the cluster out of reach of the other clusters and the PN agent.

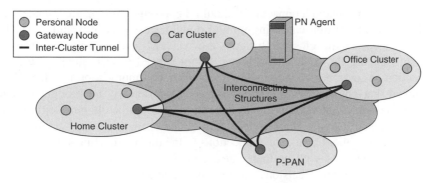

(a) Always-up inter-cluster tunneling.

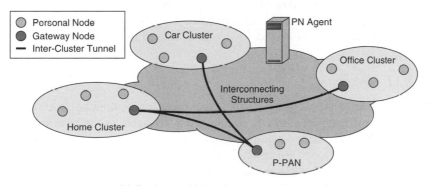

(b) On-demand inter-cluster tunneling.

Figure 6.5 Example of PN establishment.

On-demand tunneling may also establish multiple tunnels between two clusters. This requires a smart mechanism that can decide whether a tunnel should be established or not. It needs to consider the power and overhead required for maintaining each extra tunnel and weigh that against the required QoS and resilience. Hence, on-demand tunneling needs to decide whether there should be none, one, or more tunnels between two clusters and this makes the approach more complex than the always-up tunneling approach.

6.5.3 Gateway Node Coordination

When there are several gateway nodes with several available overlapping or non-overlapping connections to the interconnecting structures, there might be a need for coordination among the gateway nodes. If two gateway nodes in the same cluster have access to the same access point, there is little benefit in keeping both active at the same time. Keeping a connection active costs energy that can be saved if one of the connections is terminated. Also when the used bandwidth in or out of the cluster is small, connections can be terminated so that no idle connections are unnecessarily maintained. However, when existing connections drop out or the bandwidth requirement increases, these terminated connections should be brought up again. Hence, what is needed is a mechanism that can bring up and down access networks and their TEPs on the gateway nodes in the cluster. Bringing down a TEP can be anything from merely stopping announcing its existence so that it is never used to also stopping maintaining the connection or completely shutting down the entire communication interface. It should also be noted that connections and interfaces may periodically be brought up again to discover new possible connections and to determine whether the old ones still exist or not.

To achieve this, a gateway node coordination protocol that operates within a cluster is needed. Information about available connections to the interconnecting structures, whether they are up or down, as well as their current QoS and current load should be shared among the gateway nodes. Decisions can then be taken and again communicated back to the gateway nodes for readjustments. The protocol must understand the cost of keeping an interface up and running, the cost of maintaining a link, achievable QoS, as well as user preferences. However, it is clear that such a protocol needs many-to-many communication among the gateway nodes, similar to what was discussed in Section 5.4.4.

6.5.4 NAT Traversal

One of the main challenges of inter-cluster communication without any additional support from the interconnecting structures in addition to a PN agent is the handling of network address translators. Most access points, digital subscriber lines, or other types of Internet access that are currently offered to end customers provide IPv4 with network address translation. Consequently, to be useful today, inter-cluster tunneling must possess the capability to traverse NATs.

When the gateway nodes are not on publicly routable IP, the PN agent can assist in establishing the tunnels so that they can traverse the NATs. If only one of the TEPs for a new tunnel is behind an NAT, a message sent via the PN agent can trigger the gateway node with the TEP behind an NAT to initiate the tunnel. If both TEPs are behind the NAT, a method such as simple traversal of UDP through NATs (STUN) (Rosenberg et al. 2008) can be used. If the NATs do not allow such mechanisms, which depends on how the NATs are implemented, or there is a firewall involved, relaying of tunnel packets is an option. Traversal using relay NAT (TURN) (Rosenberg et al. 2009) is a well-known example of such an approach to NAT traversal. Relaying means that connections are established from both the gateway nodes to the PN agent, that is only out of the NATs. The traffic is then sent over these tunnels and via the PN agent.

If many gateway nodes need assistance with relaying, the PN agent may need to be powerful and have a good network connection. An alternative option is to delegate the relaying to other known relay servers not behind NATs. Such relaying servers could be other gateway nodes in the same PN that are not behind NATs, gateway nodes from other PNs, or dedicated relay servers available in the interconnecting structure provided by an operator. These servers could be organized and discovered through peer-to-peer networks similar to the way Chord (Stoica et al. 2003), Pastry (Rowstron and Druschel 2001), and Skype (Guha et al. 2006; or http://www.skype.com/) operate. If there are plenty of such relay servers in the interconnecting structures, then the effect of triangular routing can be minimized, since servers nearer the two gateway nodes can be selected.

6.5.5 Tunneling and Signaling Security

It is obvious that data traffic that crosses an external interconnecting structure needs to be protected. This includes encryption of the entire data packets, integrity protection against unauthorized alterations of the data packets, and mechanisms against replay and denial-of-service (DoS) attacks. When two gateway nodes establish at least one tunnel between themselves, they negotiate session keys. This negotiation should be protected using keys established during the personalization step. The session key should not be associated with any particular TEP or CoA. Instead, they should be associated with the gateway nodes' PN-internal addresses. This enables the establishment of a new tunnel using different TEPs without renegotiating the session keys, and this is important for timely handovers when TEPs disappear. Each packet needs to contain the encapsulated packet with the PN-internal IP header, including information such as the PN-internal addresses of the destination and source gateway nodes as well as a packet counter. The additional information is required for successful protection against replay attacks and requires both gateway nodes to keep a short list of already received packets. It should be noted that nothing of this is entirely new since the mechanisms used by IPsec in HIP (Jokela et al. 2008) are almost identical.

Not only data traffic needs protection; so also does signaling traffic. Signaling between two gateway nodes can be protected with the same mechanisms as for the data traffic. However, signaling with the PN agent must also be protected. One option is to also personalize the PN agent or make sure that the personalization step involves establishing security keys also with the PN agent. Then these keys can be used in the same way as above to protect the gateway node to PN agent signaling traffic.

6.5.6 Current Tunneling Protocols

Currently, no tunneling protocol provides both mobility support and network to network communication. Many protocols provide one or the other and a protocol for inter-cluster communication would not be very different. However, a new protocol or an extension to an existing protocol is required.

6.6 Inter-Cluster Routing

When using an always-up tunneling approach, standard routing protocols become an option. An ad hoc routing protocol, such as DYMO (Chakeres and Perkins 2009) or OLSR (Clausen et al. 2009b), is advisable due to the mobile nature of PNs and the flat addressing structure. However, when using on-demand tunneling, the routing protocol must be capable of bringing up inter-cluster tunnels when needed. In either case, the special topology of a PN, which consists of clusters with tunnels between them, requires a more tailored approach. For instance, there is very little benefit in sending full topology information between the clusters. This will only add overhead to the perhaps very limited access connections. Instead, it is better with a scheme that communicates minimal information over the inter-cluster tunnels, but is still able to achieve good routes. In the following subsections, we will describe such a routing protocol. It involves the PN agent and is suitable for both the always-up and on-demand tunneling approaches.

6.6.1 PN Agent-Based Routing

The active TEPs in a cluster are determined by the gateway node coordination protocol outlined in Section 6.5.3. However, whether to use these TEPs and establish tunnels must be determined by the routing mechanism. To achieve this, a gateway node needs to know which cluster a certain node is in and, of course, which TEPs can be used to establish a tunnel to that cluster. One approach is to let the PN agent also know the member node list of each cluster so that other gateway nodes can inquire about this as well.

Consequently, every gateway node informs the PN agent about the member nodes in its cluster in addition to its active TEPs. The list can be retrieved from the intra-cluster routing protocol if a table-driven protocol (e.g. OLSR) is used. If a reactive protocol is used, there might be a need for a special mechanism that discovers the nodes in a cluster. In either case, the PN agent should constantly be updated so that changes are propagated to the rest of the PN when needed.

For each destination node, the PN agent knows which gateway nodes can be used and which active TEPs those gateway nodes have. When a gateway node needs to send a packet to a personal node outside its own cluster, it sends an inquiry to the PN agent. In response, it will get a list of gateway nodes and their TEPs that can be used. It selects the best TEP, establishes a tunnel from its own best TEP to that TEP, and then sends the packet across. At the same time, it subscribes to the PN agent for updates related to the remote node. That is, it will be informed of any updates regarding any gateway node in the same cluster as the node so that proper actions can be taken when necessary, such as switching to a better TEP if one becomes available.

It is worth noting that the PN agent does not really need to know exactly which clusters exist. Since clusters merge and split, that would be an extra unnecessary burden to track. Ultimately, only a list of gateway nodes and their TEPs that can be used to communicate with a particular remote personal node is needed.

The next question concerns the integration between the PN agent-based PN routing and the intra-cluster routing protocol. Since it makes no difference, except in QoS, whether packets are sent via one gateway node or another, it is up to the intra-cluster routing protocol to decide which gateway node to use. OLSR, for example, has a feature called attached network. It can be used to advertise network prefixes within an ad hoc network and the gateway nodes inside the cluster. Furthermore, using the quality of the links within the cluster in combination with the quality of the available connections to the interconnecting structures should result in the best possible path at all times if we assume the bandwidth in the interconnecting structures is more than sufficient.

Let us clarify with an example. In Figure 6.6, assuming that node 1 wants to send a packet to node 7, the intra-cluster routing protocol of the cluster on the left may choose gateway node 2, since it may conclude that path A is better than path B. Gateway node 2 sends an inquiry to the PN agent about possible connections to node 7, which results in the gateway nodes 4, 8, 9 and their TEPs C, D, and E. Among the results, it selects which remote gateway node and TEP to use. This can be based solely on the QoS information about the TEP, which is the simplest solution. If so, it only picks the TEP with the best QoS parameters that can be used to connect to the destination cluster, for example, TEP C on gateway node 4.

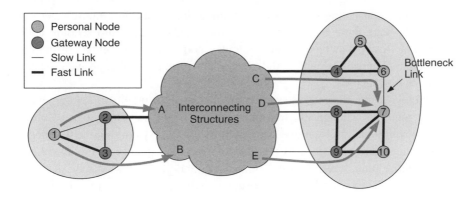

Figure 6.6   Cluster with bottleneck link.

However, there are situations where the quality of the access network may not be sufficient. Examples include cases where there is a poor quality link in a cluster. For instance in Figure 6.6, nodes 4, 5, and 6 are best connected via gateway node 4, while nodes 7, 8, and 9 are best connected using gateway node 8 or 9 due to the poor bottleneck link between nodes 6 and 7. To detect this at gateway node 2, some information about the intra-cluster routing is needed, such as the expected QoS between the gateway node and the final destination node. Gateway node 2 can then use this information in combination with the TEP QoS to decide which TEP to use. It can see that path D is better than paths C and E in Figure 6.6. However, this requires the exchange of extra information and may lead to frequent and unnecessary routing changes if the link qualities fluctuate. Hence, this should be used with care so that a reasonable level of routing stability can be maintained.

6.6.2   Tunnel Quality Assessment

To achieve good end-to-end communication quality within a PN, it is necessary to monitor the quality of available access connections as well as the quality of the entire tunnels. Usually the access connections are the bottlenecks, which means that those are the most important to monitor.

Fortunately, most access network technologies, including IEEE 802.11 (Calhoun and O'Hara 2005; IEEE 1999), IEEE 802.16e (IEEE 2006a), and UMTS already do monitor the link quality. It is part of the handover mechanism, and in those cases it is advisable to use those measurements. However, it must be remembered that those measurements are designed to choose the best access point (or base station) and hence rarely provide any further details on the quality. For proper path selection, we also need to know things like the expected transmission count (ETX) (De Couto et al. 2003a) and the data rate.

Work has been done in this area under the term 'vertical handover' or 'media independent handover' (MIH). There is currently also a standardization effort in this direction under the IEEE 802.21 working group (http://www.ieee802.org/21/). One of the aims of that work is to enable mobile terminals to make an informed selection of the best access network. The solutions are based on enabling relevant information from the lower layers and from the interconnecting structures in a standardized way. PN inter-cluster communication can leverage these technologies, but also needs to take into account additional aspects, since a cluster may have several gateway nodes. The capabilities of the gateway nodes and their connectivity within the cluster are examples of additional aspects that should also be considered.

For technologies that do not yet support any access link quality measurements, other methods are required. Though not the most effective way, network layer solutions can achieve some additional information about the link quality. By pinging the first hop access router or another gateway node, it is possible to detect the quality of the access network used and an entire tunnel, respectively. Both delay and packet loss can be detected. The main drawback is the overhead created by such a technique and the difficulty in detecting the link throughput.

When significant changes are detected in the quality, relevant nodes must be informed, such as the PN agent and gateway nodes to which there are active tunnels. However, sending minor quality updates creates a lot of unnecessary overhead without any real benefit. The latest quality information can always be included in the periodic updates that are transmitted anyway. Only when a significant quality drop is detected on a link that is currently used by traffic with strict quality requirements can an extra update message be justified. Otherwise, it should be enough to wait until the next periodic update.

6.6.3 PN-Wide Broadcasting

For some PN applications, there is also a need for PN-wide broadcasting in addition to unicast routing and cluster-wide broadcasting. Examples include revocation of security keys when a personal node is compromised (see Chapter 9), configuration updates affecting the whole PN, and service discovery.

The main issue with PN-wide broadcasting is to avoid sending the broadcast message over the interconnecting structures more times than necessary. In a cluster with more than one gateway node, there must be an agreement about which gateway node will forward the message where. One simple solution is for a personal node to first do a cluster-wide broadcasting and in parallel send the message with unicast to the gateway node with the highest quality TEP for broadcasting to the other clusters. Ultimately, this could be incorporated in the same packet. In this way, only that gateway node will take actions, thus avoiding several gateway nodes doing the same job.

Furthermore, the best TEP in a cluster may still not be fast enough to send the message to every remote cluster. Sending the same message several times may cause too much delay and consume too many resources. Further, the gateway node may not have established tunnels to all the remote clusters. Instead, sending broadcast messages via the PN agent is another good option, since it has active sessions with all active gateway nodes. Either the PN agent can do the entire job of sending the message to all other remote clusters or the gateway node and PN agent can do it together. The gateway node can deliver the packet over its active tunnels and then instruct the PN agent to transfer the packet to the remaining clusters on its behalf. Figure 6.7 shows this process assuming

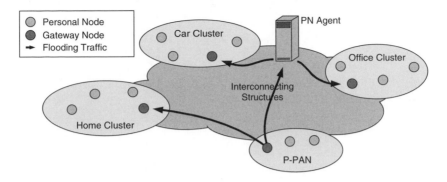

Figure 6.7   PN-wide broadcasting.

that the P-PAN cluster only has an active tunnel with the home cluster. The quality of the TEPs involved should determine how much the gateway node should do itself and how much the PN agent should do.

6.7   Summary

This chapter has focused on communication between a person's clusters. To implement this, secure intra-cluster tunnels are established and maintained. The purposes of the tunnels are to protect the intra-PN traffic as well as transparently handle cluster mobility. That is, the encrypted inter-cluster tunnels are dynamic and updated when a cluster changes its attachment point to the interconnecting structures.

Depending on the amount of infrastructure support that can be expected, we investigated several approaches to inter-cluster communication. A good option is to rely on a server located somewhere in the interconnecting structures. This server, which we refer to as a PN agent, is constantly updated with the current locations of the clusters and how they can be contacted. When tunnels must be established or updated due to mobility, the gateway nodes in the clusters may query the PN agent for assistance. Enough information is then shared to effectively establish or maintain the tunnels.

Other approaches were also investigated, such as a solution whereby access routers in the interconnecting structures can support the mobile clusters with PN-specific functionality. We also sketched a solution whereby no extra support from the infrastructure can be expected, including entities such as PN agents. For such scenarios, a PN is best implemented using peer-to-peer technology between the gateway nodes. We also discussed the issues of NAT traversal and security.

Then we looked at routing over the inter-cluster tunnels. A routing approach integrated with the dynamic tunneling was proposed. It makes use of the PN agent; the cluster node member lists are also communicated to the PN agent. When a gateway node needs to transmit a packet to a personal node not in its cluster, it may query the PN agent to find out which tunnel to use or whether a new tunnel is needed. The benefits of this approach are mainly lower overhead and the ability to establish and maintain tunnels on demand. We also discussed how to implement PN-wide broadcasting in an optimal way.

7

Foreign Communication

In the PN concept, nodes are divided into personal nodes and foreign nodes, based on which nodes the user decides to personalize. In the previous chapters, we have only discussed communication among the personal nodes themselves. Communication between a personal node and a foreign node, which we call *foreign communication*, is obviously also required. For many applications, a PN needs to interact with other PNs as well as PN-unaware devices. This includes using services from other PNs as well as offering them services. Whenever access to the Internet exists, a PN must be able to communicate with any Internet host in order to surf the web, read e-mails, etc. It must also be possible to locate remote PNs when their locations are unknown and initiate communication with them. Hence, we need to enable foreign communication, and that is the topic of this chapter.

Jane frequently uses her PN to communicate with others. When she arrives at a client, she can connect to devices present there. In many cases, she can connect to a projector or a wall-mounted display and show information about and pictures of her products to the clients. The system is set up in such a way that she only needs to locate the correct document in her PN using one of her devices and click on the 'display on big screen' button. The system automatically discovers local display options and allows Jane to choose the appropriate one. However, using location information, the system is able to immediately suggest the right option most of the time.

Foreign communication is a topic very specific to PNs. The way PN networking has been defined makes it necessary to also define solutions for foreign communication. However, the foreign communication problem can be divided into two parts. One part concerns the mechanisms within the PN, and those mechanisms are tightly coupled and specific to the PNs. The other part concerns the protocols between the edge of the PN and the foreign nodes. In the latter, existing standards are crucial. We cannot expect every foreign node to understand one or a few specific protocols. Instead, we must build on widely adopted protocols.

This chapter is structured as follows. Section 7.1 introduces the most important requirements for foreign communication. In Section 7.2, we examine how to establish foreign communication. We go on to investigate how to connect the PN-internal networking with external networks in Section 7.3, and then look at mobility aspects in Section 7.4. Finally, we summarize the chapter in Section 7.5.

7.1 Requirements for Foreign Communication

Foreign communication requires special functionality at the gateway nodes (Jacobsson et al. 2006; MAGNET Beyond 2007). It is the gateway nodes that will connect to the foreign nodes. They need to bridge the mechanisms used within the PN with the mechanisms used outside. Figure 7.1 shows two examples of foreign communication: one with a foreign node using direct communication (a common communication domain) with one of the gateway nodes in the cluster, and one with a foreign node connected through an interconnecting structure. In both cases, foreign communication must be supported and a gateway node is involved in the establishment of the end-to-end communication. Foreign communication may also need to use other types of networks to reach the destination device, such as multi-hop ad hoc networks. In either case, the gateway nodes need to understand and participate in the mechanisms of the external networks. Furthermore, they must be able to accept and handle connections initiated by foreign nodes as well.

For foreign communication, it is crucial that the following requirements are met:

(i) Communication between applications on personal nodes and applications on foreign nodes must be possible. Both PN-aware and PN-unaware foreign nodes must be supported without requiring special functionality in the PN-unaware foreign nodes. Both foreign nodes and personal nodes must be able to initiate such communication. However, it should be noted that this end-to-end communication is not necessarily provided at the network level.

(ii) Both personal and foreign nodes can be mobile. Hence, mobility must be supported when either the personal nodes or the foreign nodes roam. Mobility of the personal nodes should also be handled when communicating with mobility-unaware foreign nodes. Furthermore, the system should be able to find and select the best possible communication path at all times.

(iii) Just like the intra-PN communication, all foreign communication must be self-organized. Foreign communication must be established and maintained without

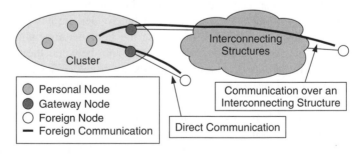

Figure 7.1 Types of foreign communication.

requiring the user's involvement. Furthermore, it should be able to automatically discover communication opportunities that the user may benefit from, both new connection possibilities and new interesting foreign nodes and their applications.

(iv) Foreign communication must work well even when the number of interactions between PNs and between PNs and PN-unaware nodes is large. Any solution must assist the user in managing these interactions.

(v) Since foreign communication involves PNs and devices of other persons and organizations, security and privacy are even more important and difficult. Despite this, efficient and easy-to-use solutions are required that can handle the security risks introduced by foreign communication.

In the remaining sections of this chapter, we will introduce and discuss options and solutions that can successfully meet all these requirements.

7.2 Setting up Communication with Foreign Nodes

For security reasons, it is important that foreign communication mechanisms remain separated from intra-PN communication mechanisms. This means that the gateway nodes need to treat foreign communication in a different way than intra-PN communication, for example, block all non-approved traffic from entering the PN. The gateway nodes must bridge the mechanisms used inside the PN and those used to communicate with the foreign nodes, since these are different and should never be combined. Foreign nodes should also be properly authenticated before any communication is initiated. This topic will be covered in Section 9.3.

As shown in Figure 7.1, the gateway node may connect with the foreign nodes in several different ways. Each way has its own mechanism and hence requires different approaches:

Direct communication. At the connectivity level, the gateway node must of course share a common communication domain with the foreign node for direct communication. At the network layer, it needs to establish a network connection to the foreign node, which may be an ad hoc and temporary connection. Temporary link local addresses can be used in the foreign communication as long as they are unique among the communicating peers. If available, another option is to use an already deployed network (such as a WLAN hotspot), where addresses usually are assigned automatically, for example, by DHCP (Droms 1997).

Communication over interconnecting structures. If a personal node wishes to communicate with a foreign node that is connected through an interconnecting structure, the gateway node that links the cluster to the interconnecting structure needs to bridge the PN-internal mechanisms with the mechanisms used in that interconnecting structure. In this case, there is also the possibility of using the PN agent as the bridge.

Communication over other network types. Foreign communication may also need to use other types of networks, such as multi-hop ad hoc networks. The gateway node needs to understand and participate in the mechanisms of the external network. Consequently, gateway nodes may need to support several different network types.

It is also important to be able to switch between the different communication approaches when better alternatives arise or existing ones disappear (Carter et al. 2003). We return to this topic in Section 7.4 when we discuss mobility.

7.2.1 Foreign Node Discovery

The first step in establishing foreign communication is to discover potential foreign nodes to communicate with. This step, called node discovery, is the task of the potential gateway nodes, as they are the only nodes that can interact directly with the external networks. Gateway nodes should keep a list of foreign nodes (or networks) so that foreign communication can be established when and if required. The gateway nodes may inform the other nodes in the cluster, or the whole PN, about the foreign nodes (or foreign PNs) by broadcasting this within the PN as part of the routing protocol. At the service level, the gateway node may also discover services on local foreign nodes and advertise these within its cluster. This can be done by populating the service management node (SMN) in its cluster (Ghader et al. 2006; MAGNET 2004b) as discussed in Section 4.3.3 and Chapter 8. If the external network is PN-aware and has an SMN, it may interact with it.

To enable communication to remote foreign nodes, the gateway nodes advertise their current interconnecting structure connections. This advertisement can be done as a simple default gateway (or network prefix) within the cluster. In this way, every node in the cluster knows which gateway nodes can be used to connect to the interconnecting structure.

Once a foreign node has been discovered and a personal node wishes to communicate with it, a gateway node must first be selected. If the personal node itself is a gateway node and has a link to the foreign node, it should choose itself instead of relying on other nodes. If this is not possible, or not desirable (e.g. its own connection is limited or costly), it may choose to use another gateway node instead. It may be possible to choose from more than one gateway node. In some cases, as in Figure 7.2, a direct connection (A) and several infrastructure-based connections (B–D) are possible at the same time. The node needs to carefully select one of them, since the traffic between the two end nodes has to go through the selected gateway node. In many cases, it is very hard to change gateway node without tearing down the connection and establishing a new one.

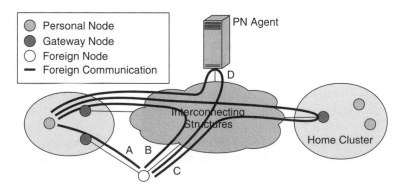

Figure 7.2 Four gateway node options.

States need to be transferred between the two gateway nodes and the foreign node may not support mobility. A last option is to use the inter-cluster tunnels and choose a stable gateway node (e.g. C) in a stable cluster or the PN agent (D). In that case, mobility of the personal node is handled internally by the PN itself as described in Chapter 6, but at the same time the routes are not optimal.

7.2.2 Accepting Connections from Foreign Nodes

In some scenarios, it is interesting to consider the case where a foreign node wants to initiate a connection to the PN. If there is a direct connection, the foreign node can just initiate a connection to the present gateway node, which needs to handle it and establish an end-to-end session with one of the personal nodes within the PN.

When no direct connection exists, the foreign node needs to use the interconnecting structure. However, the nodes of a PN can be mobile and change their point of attachment and therefore also their addresses used in the interconnecting structure. The only entity that does not change address is the PN agent. The PN agent is therefore an excellent point of contact for foreign nodes that wish to establish communication with a PN. It is only necessary to remember the address of the PN agent to be able to initiate connections with that PN. To simplify the process even further, the address of the PN agent can be given a name that can be resolved through DNS. The PN agent will know the location of all clusters in its PN and can tunnel the packets to the final destination within the PN. At the same time, it will bridge the interconnecting structure and the intra-PN mechanisms.

Jane handles all her telephone calls with her PN. She has connected both a telephone number and an SIP address to her PN. Both of them actually lead to the PN agent, but that is something Jane is unaware of. When a call is made to Jane's PN, it diverts to a device near her. This is done intelligently based on a combination of presence detection, the most recently used device, as well as user preferences configured by Jane, based on the time of day.

7.3 Bridging Inside and Outside Protocols

After a foreign node has been discovered, it is time to establish data communication. The gateway node participates in the external network in which the foreign node exists. If the gateway node itself wishes to communicate with the foreign node, nothing extra is required. The gateway node uses the mechanisms specified by the external network and acts as a normal node in the external network. However, if another personal node in the PN wants to communicate with the foreign node, then the gateway node needs to bridge the two network types.

There are two different ways a gateway node can connect the PN to the external network. End-to-end IP connectivity can be established between the two communicating endpoints using network address translation in the gateway node. An alternative is a *service proxy* that bridges the intra-PN and the external network. Section 7.3.1 discusses the network abstraction level solution using NATs, while Section 7.3.2 covers the service proxy solution at the service abstraction level.

7.3.1 At the Network Abstraction Level

In this approach, the source personal node needs to send packets to the destination foreign node such that they go through the selected gateway node. To do this, the source node cannot just put the address of the destination node as the destination address of the packet. The intra-PN routing mechanism should not need to bother with every foreign node address as they may overload the intra-PN routing tables. Further, if two gateway nodes have a path to the same foreign node, it is not guaranteed that the packets will go through the selected gateway node all the time. The routing protocol may decide to switch gateway node at any time without prior warning. Finally, the address space may overlap and two different foreign nodes may use the same address. This can happen if the foreign nodes use RFC1918 addresses (Rekhter et al. 1996), addresses with link local scope (Hinden and Deering 2006), or if some nodes want to sabotage the PN operation by deliberately choosing non-unique addresses. This leaves three options for the personal node to make sure that the packets go through the selected gateway node:

Tunneling. An IP-in-IP tunnel is established between the personal node and the selected gateway node. The endpoints of this tunnel are intra-PN addresses while the packets in the tunnel use the foreign node address as the destination. To send packets in the other direction, from the gateway node to the personal node, no special functionality is necessary.

Source routing. The intra-PN address of the gateway node is used as the destination address and the final destination, the address of the foreign node, is placed in an option field of the IP header, for example, using a Type 0 routing header in IPv6 (Deering and Hinden 1998). In principle, this scheme and the tunneling scheme are the same. However, source routing creates slightly less overhead.

Address aliases. We cannot use the addresses of the foreign nodes as mentioned above, but we can inject address aliases into the intra-PN routing protocol. If each foreign node and gateway node pair has a unique address alias within the PN, then a personal node can select both the foreign node and the gateway node to be used, hence the control remains with the source node. To minimize the burden on the intra-PN routing protocol, only aliases of (foreign node, gateway node) pairs that are in use should be injected.

While the personal node connects to the gateway node, the gateway node also needs to set up a path through the external network to the foreign node and install a state for the address translation so that intra-PN addresses can be translated to the addresses used outside. When all this is in place, traffic can start to flow. The personal node sends the packets to the selected gateway node, which will remove all PN-specific headers, encryptions, etc. Network addresses are then translated before being forwarded to the foreign node. To the applications, it will look like there is an end-to-end IP connection between the two end nodes. Figure 7.3 shows an example of communication between a node in a PN and a PN-unaware foreign node through a gateway node according to this scheme.

The address translation can be done with standard network address port translation (NAPT) technology, usually known as just network address translation (Srisuresh and Egevang 2001). NATs are usually used to extend the limited address space in IPv4, but

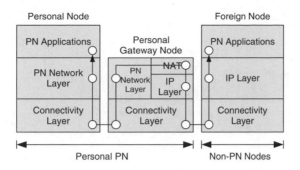

Figure 7.3   Foreign communication with a PN-unaware node.

can also be used to hide internal address spaces used on private networks, such as PNs. The address space within a PN is to be used within the PN and is not globally unique. Because of this, network addresses must be translated and mapped when packets flow in and out of a PN and this is a typical task of NATs. Hence, standard NAT technology will work.

The next requirement concerns accepting incoming connections from foreign nodes; it is actually not very different from the outgoing connections. First, the gateway nodes must somehow make sure that the PN is known to the outside world, for instance by advertising itself in the local vicinity. However, no PN-internal addresses should be advertised outside the PN because of privacy reasons and the fact that PN-internal addresses are not unique. Instead, the gateway nodes advertise their own external addresses used within the external networks plus the services that the cluster or PN can offer. Then, as requests to establish communication arrive at the gateway nodes, they need to determine whether to accept the connection or not. If the connection is approved, an appropriate personal node is chosen to become the end node. Higher layer information may determine this, based on what type of connection or what service is requested.

From here onward, there is no difference between this case and the case where a personal node initiated the connection. An address translation state needs to be installed in the gateway node that translates the network addresses. When packets arrive at the gateway node, it adds the necessary headers and encryption before the packets enter the PN and are forwarded to their final destinations. All the rest remains the same, except that the foreign node will select the gateway node in this case.

The case where two PNs want to communicate with each other is nothing more than the combination of the two scenarios mentioned above. Figure 7.4 illustrates this.

7.3.2   At the Service Abstraction Level

In many cases, it is actually not necessary to have end-to-end IP connectivity between the end nodes. Another option is possible if we can assume that everything is provided as services using a common service provisioning framework such as SOAP (W3C 2007). Then the gateway nodes can use service proxies to relay services outside the cluster into the PN and vice versa. If a gateway node discovers a printing service outside its cluster, it can offer that service to the nodes inside its cluster by starting a service proxy. A client inside the cluster can then use the outside printing service by using the service proxy

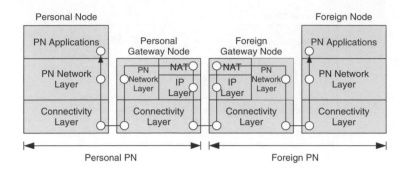

Figure 7.4   Foreign communication with another PN.

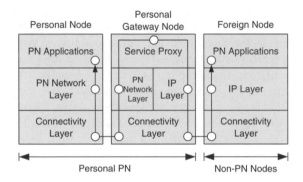

Figure 7.5   Foreign communication at the service level with a PN-unaware node.

in the gateway node. The service proxy forwards the service calls to the external printer service and sends back the replies from the printer to the client. In other words, it acts as a server to the client application and as a client to the printing service. The main purpose of the service proxy is to relay traffic in and out of the PN, which is necessary if this is not provided by the network level. Figure 7.5 shows this case. Note that there is no longer a need for NAT at the network level. Instead, there are two separate network connections; one from the client application on the personal node to the service proxy on the gateway node and one from the service proxy to the server application on the foreign node. The service proxy just connects the two, but at the application layer.

The same solution can be used when a PN wants to offer services to the outside world. The gateway nodes or the PN agent can export these services by means of similar service proxies. Such a service proxy exports the services from the personal nodes and makes them available for foreign nodes. This solution may even be better than the network level solution from the point of view of security, since the gateway nodes and/or the PN agent can control which services are exported and which ones are not. A finer granularity of access control is also possible, since not all parts of the service interface might be exported.

If we assume a service framework based on SMNs, as mentioned in Section 4.3.3, Chapter 8, and the MAGNET architecture (Ghader et al. 2006; MAGNET 2004b), then there is a SMN elected within the cluster that manages the services, clients, and service

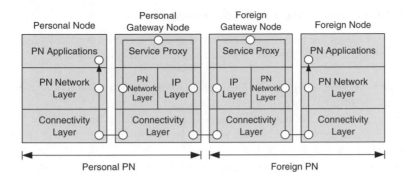

Figure 7.6   Foreign communication at the service level with another PN.

sessions. The SMN should also manage and coordinate the service proxies on the gateway nodes and instruct them as to what services to export or import, and to what degree. The SMN may also control which gateway node a particular service proxy for a particular service should run on. The fact that the gateway nodes and the SMN provide this functionality also means that simpler devices can provide services to the outside world, since they would not need to bother about access control, authentication, etc.

Figure 7.6 shows the case where a personal node in a PN uses a service offered by a foreign node in another PN. The foreign gateway node exports a service from one of the nodes in its cluster. Then the personal gateway node detects this service and offers it to the personal node using a service proxy. In this case, there is a difference between the service proxy that relays external services into the PN and the service proxy that exports services to the outside world. The former tries to secure the service usage by using certificate servers, or other similar methods, while the latter implements access control and authorization mechanisms. Both service proxies could operate under the control of their respective SMNs in the two PNs.

7.3.3   Network versus Service Abstraction Level Approach

A drawback of the service proxy approach is that we now have two connections instead of one at the network level. One connection is established between the client and the service proxy and another one between the service proxy and the service. If two PNs are communicating, there might be a third one between the two proxies running on each of the gateway nodes as shown in Figure 7.6, and this may adversely affect the QoS of the end-to-end communication.

It is of course possible to use both the network level and service level solutions at the same time. QoS-sensitive communication might use the network level mechanisms to establish end-to-end IP connectivity between the end nodes, while others may use the service proxy approach. Another option is to use the network level mechanisms for personal nodes using services outside the PN, while a service proxy is used when the same personal node exports services to the outside. In this way, we will avoid having two service proxies and increase the possibilities for access control of the exported services.

7.4   Mobility and Gateway Node Handover

Since many PN nodes are mobile, it is natural that also foreign communication paths need to support mobility. There are several reasons why mobility is required:

(i) The gateway node switches its point of attachment to the interconnecting structure, requiring a different care-of address (CoA).
(ii) Direct communication between a personal node and a foreign node is no longer possible due to mobility. Consequently, a switch to interconnecting structure-based communication may be required.
(iii) When direct communication becomes possible, it is usually better to switch from a connection via an interconnecting structure to a direct connection. In most cases, a direct connection can offer better bandwidth, better QoS, and lower cost.
(iv) The selected gateway node becomes unavailable or loses its connection to the foreign node (or the interconnecting structure) and another gateway node must be used.
(v) The foreign node might be mobile. However, in this case, it can be assumed that it has its own support for mobility.

These examples demonstrate the importance of good mobility solutions for foreign communication, as we would like ongoing sessions to proceed without interruption.

There are at least two solutions for foreign communication mobility. The first one is very simple, but non-optimal. It relies on always using interconnecting structures and routing the traffic through the PN agent, which is a fixed node with a fixed address. The second solution is more complex but achieves much better routing, since traffic is routed through the most appropriate gateway node. Mobility of the gateway node (terminal mobility) as well as a switch to another gateway node must be supported in this case.

7.4.1   Always Using the PN Agent

The main advantage of always routing foreign communication via the PN agent is its simplicity. All foreign communication goes through the PN agent, which means that only the PN agent needs to implement the bridging as shown in Figure 7.7. The PN agent never

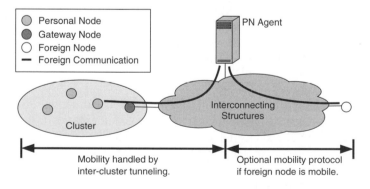

Figure 7.7   Always using the PN agent for foreign communication.

changes its address so there is no need for any external mobility solution between the PN agent and the foreign nodes unless the foreign node is mobile. In addition, the intra-PN routing will be very simple. Packets with a foreign node destination are forwarded to any gateway node in the cluster and then over the interconnecting structure to the PN agent. A default route within the PN can achieve this. Hence, there is no need for source routing or tunneling to make sure the packets arrive at a particular gateway node.

Obviously, this solution has many limitations. To route all foreign traffic (both directions) through the PN agent leads to non-optimal routing and a potential bottleneck. Furthermore, direct communication to a foreign node is not at all possible. On the other hand, this solution handles mobility of the PN clusters also when communicating with foreign nodes. The link between the PN agent and foreign node remains stable and does not change unless the foreign node is mobile. All mobility of the personal nodes takes place within the PN and is handled by the dynamic tunneling provided by the PN, which can be extended to include the PN agent. The cluster gateway nodes inform the PN agent about their CoAs, including changes thereof. This enables the personal nodes to be mobile while communicating with foreign nodes, and this only at the cost of non-optimal routing on the interconnecting structure side.

This solution is similar to Mobile IPv4, when using reverse tunneling as described in Montenegro (2001), except that foreign agents are never used in our case. While it would be possible to actually use Mobile IP, it would mean using two mobility solutions in parallel, which is of course unnecessary. Hence, it is better to rely on the mobility mechanisms already provided by the PN.

There is one more important reason why this solution is good, and that is privacy. If the traffic goes directly from the gateway node to the foreign node, the CoA of the gateway node will be known to the foreign node and this address may reveal the user's current location. If the traffic goes via the PN agent, this address will be hidden from the foreign node, thus guarding the location privacy of the user.

7.4.2 Using the Optimal Gateway Node

To enable mobility with optimal routing for foreign communication, two problems must be addressed. First, address changes and multi-homing at the gateway node must be handled. Second, support for switching between two gateway nodes (or the PN agent) must also be available. To sustain the ongoing connections, both require some sort of mobility support between the gateway node and the foreign node as shown in Figure 7.8. Hence, a common external mobility protocol is needed. In addition to this, a gateway node handover also requires the two gateway nodes to exchange state information.

Consequently, there is a need for an intra-PN protocol that can communicate the intention to change gateway node and then transfer these states between the two gateway nodes. The protocol should preferably be able to act before the old gateway node loses its connectivity or becomes unavailable. The protocol must also trigger the external mobility protocol to take appropriate actions.

It cannot be assumed that foreign nodes can handle a sudden change of address at the gateway node without mobility support. Most current IP nodes on the Internet or elsewhere cannot handle such changes without losing the connection. A widely deployed mobility standard is required; otherwise, the chances of a foreign node actually implementing a

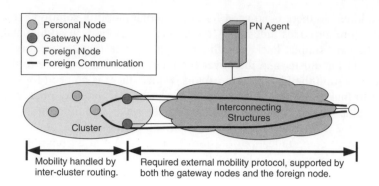

Figure 7.8   Using the optimal gateway node for foreign communication.

proper solution are very slim. With this in mind, there are only a few options that can be used to handle mobility between the gateway node and the foreign node:

Mobile IPv4 (Perkins 2002). This is a well-established protocol for mobility on IPv4 networks, which we introduced in Section 6.2. The PN agent can act as home agent (HA) and the foreign node can use IPv4 as usual without any modifications. Whenever a gateway node changes address or another gateway node is selected, the HA at the PN agent is informed. However, the foreign node cannot be informed, which is a limitation of Mobile IPv4 as it is now being standardized. Consequently, the foreign node will always send its packets via the HA (PN agent), which is still non-optimal. Only packets from the personal node to the foreign node will take the direct path. Furthermore, Mobile IPv4 is not able to switch to direct local communication when such possibilities exist. Instead, all traffic has to go via interconnecting structures. Hence, the benefits of this option, compared to always routing through the PN agent, are limited.

Mobile IPv6 (Johnson et al. 2004). The biggest difference between Mobile IPv4 and Mobile IPv6 is in the use of binding updates, which are the messages sent to the HA to update the CoA. In Mobile IPv6, binding updates are also sent to the foreign node. If the foreign node implements IPv6 and Mobile IPv6, it can directly send its packet to the gateway node instead of via the HA (i.e. PN agent). Otherwise, the two protocols work in the same way and have the same limitations.

Host Identity Protocol (HIP) (Moskowitz et al. 2008). HIP is also able to handle mobility (Nikander et al. 2008), though it currently seems that it too, lacks support for local direct communication. However, HIP is not yet a standard and still has a long way to go before becoming one. It is therefore unlikely that we can expect foreign nodes to implement HIP in the near future. However, if the foreign node supports HIP, then this could be a good choice.

Transport and application layer mobility protocols. As explained in Section 6.2, many proposals have been made for mobility handling at the transport layer (Ansari and Sathyanath 2007; Maltz and Bhagwat 1998; Snoeren and Balakrishnan 2000; Zandy and Miller 2002) and at the application layer (Schulzrinne and Wedlund 2000). However, all of them have the same problem as HIP; none is a standard yet and none has any real deployment. Should any of them take off and become widely deployed, then they would all be good candidates for mobility support when bridging at the service level.

Contact Networking (Carter et al. 2003). Currently, most mobility protocols only focus on global and remote communication over interconnecting structures. Contact Networking (Carter et al. 2003), on the other hand, also supports local direct communication. Contact Networking tries to use direct communication if possible. When that is not possible, it falls back on the interconnecting structures using techniques similar to Mobile IP. If direct local communication later becomes available again, it switches back. Contact Networking combines neighbor discovery, automatic addressing, link break detection, routing, vertical handover, mobility, and naming into one single solution. However, Contact Networks requires all corresponding nodes also to be aware of Contact Networking, which means that a wide deployment is first needed. Unfortunately, there is currently no standardization or development effort going on in the direction of Contact Networking. Further limitations include the fact that there is no security and of course the inability to handle an entire PN instead of a single device. Otherwise, Contact Networking is perhaps the best option for foreign communication mobility if it becomes a widely deployed standard.

In an ideal world, we conclude that a solution based on a combination of HIP and Contact Networking is probably the best solution. Such a solution should be able to handle both security and mobility across both direct connections as well as interconnecting structures for all types of foreign communication. However, this requires most foreign nodes to support those protocols, which is currently not the case.

If two PNs are communicating, there is one more possibility. It would be possible to deploy special functionality at the gateway nodes that handles mobility between the PNs. In the event that any of the gateway nodes needs to change network address, a special inter-PN mechanism could be used to maintain the connection. They could exchange the addresses of their PN agents to fall back on, in case the current communication link breaks. In addition, the address changes should be communicated directly between the foreign gateway nodes in the same way as in the protocols above.

Another important aspect of gateway node mobility is the handover delay. Coordination among the personal nodes and the gateway nodes, selection of a new gateway node, transfer of states, and the operations of the external mobility protocol all introduce delay. If a connection carries real-time data, serious quality problems may arise if the time to adapt becomes too long. Acting in advance, with a 'make before break' approach in combination with very speedy operations of all mobility interactions involved, is preferable.

7.4.3 Using Service Proxies

When using service proxies instead of NAT, it is still possible to use the same solutions for mobility between the gateway node and the foreign node. There is also the possibility to use service level mobility such as SIP mobility (Schulzrinne and Wedlund 2000). Inside the PN, nothing at the network level needs to change.

The only additional problem that needs to be handled is the change of service proxy. Imagine that a gateway node running a service proxy is about to become unavailable, then a change to another gateway node with the same type of proxy is needed. This would require similar handover procedures as in the network level solution. However, the amount of state information that must be transferred may be larger and more complex. Typically, a service proxy will keep more state information related to the service itself, such as buffered data and remote procedure calls. Except for this, a transfer from one gateway node to another will work in the same way.

7.5   Summary

In this chapter, we have looked at how communication with PN-unaware devices and foreign nodes can be done. For security reasons, the gateway nodes need to treat foreign communication in a different way and block all non-approved traffic from entering the PN. Intra-PN communication mechanisms must remain separated from foreign communication mechanisms. Hence, the gateway nodes must bridge the mechanisms used inside the PN with those used to communicate with the foreign nodes.

Two different solutions were described; either end-to-end network layer connectivity is established across PN boundaries using network layer address translation or a service proxy bridges between the PN and the foreign nodes outside.

Since many PN nodes are mobile, foreign communication paths also need to support mobility. We introduced two approaches to handle this: either always sending foreign traffic via the PN agent and handling all mobility with the inter-cluster tunneling mechanisms, or using the most optimal gateway node and then using a standardized and well-adopted mobility protocol between the gateway nodes and the foreign nodes. The first option is the simplest and works with all current foreign nodes as well as PN-unaware devices. The second option is more complex and requires the foreign nodes to implement a mobility protocol, but also allows for more efficient routing.

8

Personal Network Application Support Systems

Personal networks are intended to offer a platform to the user with which it is possible to access and use all one's personal devices in an unobtrusive and secure way. Being unobtrusive means that the user should as little as possible be aware of or have to bother with the technical aspects of a PN. Neither should the service developer or application programmer have to worry more than necessary about the technicalities of the underlying platform. Indeed, the easier it is to develop new applications and services, the more applications will appear for PNs.

To make this happen, it is necessary for the PN platform to take as much of the burden as possible off the application developers. The PN should offer an application programming interface (API) that is rich enough to exploit the full potential of PNs, while at the same time hiding the details of the PN operation that are not relevant for the applications. The application developer should have at his disposal the right PN services to develop attractive applications for the user and service providers alike.

Currently, PN application and service support have not been researched and experimented with as thoroughly as communication and networking aspects. However, both the IST MAGNET (http://magnet.aau.dk/) and Freeband PNP2008 (http://pnp2008.frecband.nl/) research projects have made significant contributions to PN application and service support.

In this chapter, we discuss the achievements of these projects in this area. In Section 8.1, we address the very diverse nature of the applications that might, now and in the future, run on PNs and the support these applications need. In Section 8.2, some concepts as well as functional components that might give this support are introduced. Then, in Sections 8.3 and 8.4, we move on to discuss in more detail the work on service discovery and management and context management, respectively. In Section 8.5, we summarize the chapter.

8.1 Required PN Application Support

PNs should create the opportunity for a vast array of new applications but should support legacy applications as well. For instance, if a person has a PN encompassing his smart

Personal Networks: Wireless Networking for Personal Devices   Martin Jacobsson, Ignas Niemegeers and Sonia Heemstra de Groot
© 2010 John Wiley & Sons, Ltd

phone or laptop, it should be able to make phone calls, check e-mail, run web applications such as Twitter, or participate in a social network such as Facebook.

The merits of a PN, however, should mainly be in enabling new applications and services, and in enhancing existing ones. An example of a PN enhancement of an existing application would be that when a phone call comes in, it is automatically directed to the personal device that is most convenient for the user, given the context he is in. This might be his smart phone, a cordless home phone, a laptop, or even an audio/video entertainment system if person is relaxing at home and receives a call while enjoying multimedia entertainment. This can be realized by virtualizing the resources in a PN: the application is only aware of a virtual audio (eventually perhaps a video) communication device, which is mapped onto a real device or multiple devices by the PN based on context.

Completely new applications for PNs will be found in domains such as health and well-being, security and safety. Many of these applications will be embedded, that is, their presence may hardly be noticed by the user.

Jane is a passionate runner. She is interested in keeping track of her performance so that she can adjust her training program accordingly. For this purpose, she has purchased a PN-enabled sports watch with GPS functionality that registers and stores her heart rate and her instant location and trajectory. The watch implements a coach application for runners that monitors Jane's progress and instructs her on a suitable training program.

The PN-enabled sports watch also has a health monitoring application, which is only activated when needed. The application will never be noticed by Jane, except for letting her know that all is well or alerting her when she needs to be careful or when help is required.

People will also need to offer services to others and share resources with each other, which is, to a large extent, the topic of Chapters 7 and 10.

Before we start discussing application and service support in detail, we need to define what we mean by these terms. We consider an application to be a logical component implemented by a program running on a node. The application may be started as a direct consequence of an action of a PN user or triggered by an event in the case of an embedded application. A service, on the other hand, is considered to be a logical component implemented by a program running on a node that offers something that an application can use through a specified interface. Both applications and services may be users of other services. It is possible that an application consists of the application itself and a combination of services it uses that may be running on other personal nodes. Figure 8.1 shows such combinations and how they map onto the PN. Compare this figure with Figure 4.1.

Jane's sport watch is running her sports coach application. For the application to be usable, it needs a user interface. If nothing is available at the moment, the application cannot start. Jane can, for instance, use a small screen that she can clip to her belt or her portable music player. With the latter, voice commands are given to Jane while she is running.

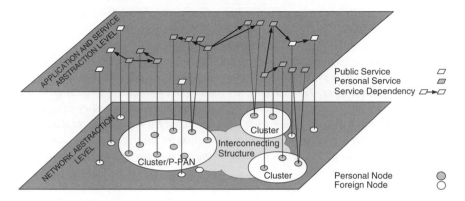

Figure 8.1   The service abstraction level view of a PN.

Normally, the coaching application would use the built-in GPS and heart-rate sensor. However, Jane has a better GPS receiver with better accuracy and her clothes have built-in health sensors that provide more and better health readings. All those sensors are PN-enabled devices and offer their sensing services to any PN application. Hence, the coaching application may use those services instead.

The coaching application also has the capability of instructing certain exercise equipment, such as a treadmill. However, those functions are not activated unless a treadmill service becomes available in Jane's PN.

These are examples of how an application can depend on services or make use of better services, should such be available. In the same way, services may themselves in turn depend on other services as shown in the top half of Figure 8.1.

Let us now take a look at specific functional features that an application and service support system should have in order to realize the full potential of PNs.

8.1.1   Naming

One very important application support system feature is naming and name resolution. The use of names is crucial in order to build a user-friendly PN. Basically anything that needs to be seen by the user should have its own name, such as nodes, PNs, services, and perhaps even clusters. The naming scheme must provide a flexible naming resolution to the PN and, at the same time, be compatible with the naming schemes used in the current infrastructure, for example, the Domain Name System (DNS) (Mockapetris 1987). For inter-PN naming solutions, the MyNet project may provide an excellent solution (Kaashoek and Morris 2006).

8.1.2   Ubiquitous Access to Services

Ideally, the use of applications and services provided by the PN to its user should no longer be tied to a particular location, but should be independent of where the user and his personal nodes are geographically located. Of course, this is all subject to the availability of sufficient connectivity between personal nodes in clusters and the inter-cluster connectivity.

> *Jane's sports coach application, which involves telemetry, real-time data analysis, and audio feedback to a person, could, provided the data streams are not excessive, be run on a remote server. This would only make the application available in places where good Internet access is available. On the other hand, the life-critical health monitoring and intervention application also provided by Jane's sports watch would require a high degree of dependability that cannot be guaranteed everywhere. The PN might even have to warn Jane when she ventures into areas where support cannot be guaranteed sufficiently.*

In order to make this possible, a PN-wide service discovery facility will be needed, which allows applications and services themselves to find what services are offered and on which nodes. This requires formalized descriptions of services, to enable applications and other services to discover them. These descriptions must specify what the service offers, how to use it, and on which node it resides.

8.1.3   Pooling of Resources

The power of a PN is that it brings to its user the capabilities of all the personal devices he has and the synergies this may yield. Of course, this is subject to the accessibility of these devices. Many of these devices will have similar or the same functionality. This opens the opportunity for the applications to pick and choose the most suitable ones. This in turn requires that a PN is able to pool its resources, such as storage and computation, so that from this pool selections can be made to support the applications. It also implies that the PN should be aware of all its resources and their availability.

For example, the cluster that is physically closest to the user himself may contain multiple devices for displaying information and interacting with the user, and there may be multiple computational and storage devices to run applications on. An application requiring a high degree of reliability might exploit these redundancies by using several gateway nodes in parallel, have backup facilities for interacting with the user and have standby computational and storage resources. On the other hand, if low energy consumption or system autonomy is of primary concern, the resource pooling should be exploited to pick those devices to run the applications that minimize overall or local energy consumption and yield maximum battery life. Similar to service discovery, pooling of resources will require a facility by which these resources can be formally described and located.

8.1.4   Gathering and Exploiting Context

A PN and its services and applications should be able to adapt their behavior according to the context. A broad and widely used definition of context was proposed in Dey (2000):

> Anything that can be used to characterize the situation of an entity. An entity is a person, place or object that is considered relevant to the interaction between a user and an application, including the user and application themselves.

Since a PN has multiple constituent parts, which can be geographically spread over a wide area, in principle even globally, the context for the PN can be very complex. It encompasses:

(i)   the state of all the resources of the PN;
(ii) the state of all the applications and services running on the PN, including the inter-actions with outside (non-PN) entities;
(iii) the geographical location of all devices involved and the user;
(iv) any other relevant information external to the PN, such as the weather forecast.

The PN should be aware of this complex context and it should be able to adapt accordingly. For example, a security-critical PN application might have to be aware of the fact that a user and some of the devices involved are in an area where the security provided by an Internet WLAN access point is insufficient and might cause a switch to a more secure cellular access. On the other hand, consider an application that is intended to monitor the health of a vulnerable person while outside a medical institution, which requires a high level of protection of all information going around in the PN. Such an application might have to drop its protection level in the event of an emergency to allow a doctor who happens to come to the rescue to extract crucial information to assist the person in need.

Applications where the system autonomy is important might want to switch to new computational and communication resources that become available, in order to conserve battery energy of some crucial devices. For instance, if a person arrives at home, many applications running on battery-powered devices might switch to devices that are plugged into the mains.

In order to support this, a context management facility is needed in the PN. It should be able to supply to any applications or system function, in principle at all levels, the relevant context information. Moreover, it should be able to collect and store this context information. Given the diversity of context, it should also be an extensible facility, able to incorporate any new type of context deemed useful.

8.1.5 Ability to Optimize and Make Tradeoffs

The PN should make the right choices and the right tradeoffs in providing services to its user in the best way possible. The availability of multiple resources with similar functionality creates the opportunity to optimize the way applications and services are offered. There may be different objectives, for instance, minimizing local or global energy consumption, maximizing system autonomy, achieving a particular level of security, offering the highest perceived quality to the user, or incurring the lowest monetary cost. These objectives may depend on the context; for example, when the user is in a public area, security may become an important concern. A particular challenge arises from the fact that multiple concurrent applications may be present in the PN, and for each of them different objectives may be appropriate. Since all of them might potentially share the same resources, this may lead to conflicts. The PN should be able to resolve those conflicts and make tradeoffs. An example is when a health support application is running and an emergency occurs. Under those circumstances, the PN should be able to give absolute priority to the concerns of the health applications and suppress other applications if needed.

In order to do the optimizations and make the tradeoffs, the PN obviously has to be aware of the state of its resources, such as their availability, energy levels, and connectivity, and of the explicit and implicit demands of the user.

8.2   Design of a PN Application Support System

Let us now take a look at how we can design a PN application support system, such that it has the features discussed in the previous section. The role of the PN application support system is twofold. The first role is to create an abstraction of the underlying PN on behalf of the application programmer, that is, provide an appropriate API. The second role is to provide middleware-like mechanisms for supporting the development and running of PN applications. Some of these mechanisms may be exposed to the application programmer, others may be hidden. Let us first discuss the abstraction of the underlying PN, followed by the mechanisms and their interrelations.

8.2.1   Abstraction for the Application Programmer

The task is to create an appropriate abstraction that, on the one hand, insulates the application programmer from the complexities of the very heterogeneous and dynamic distributed system that a PN is, and, on the other hand, gives him sufficient control to exploit the capabilities of the PN. The ultimate goal is to lower the threshold for developing PN applications, so that the development cost is low and large numbers of applications are made. The challenge is to strike a balance between what is revealed to the application programmer and what is hidden. The more is revealed, the more control the application programmer has and, in principle, the better he can exploit the PN potential, but also the higher the risk of badly engineered applications that detrimentally affect the operation of the PN. This is particularly relevant because the intent is that many different parties will be developing applications for PNs; these should ideally not interfere with each other and share the PN resources in a fair way.

What in particular should the API reveal or what should it hide, and why? This is a difficult question to answer, requiring not only further research, but also in particular experience in application development for PNs and the running of PN applications. However, focusing on the strengths of the PN concept, we can make the following observations:

(i) The application programmer should be able to specify the QoS expected from the PN. There may be many personal resources with redundant functionality that are, in principle, available to support an application. It is a major asset of a PN that choices can be made that result in very different ways in which an application is supported. We feel that the API should allow the communication of what quality is required under what circumstances. The API should give back information on the quality that can be and is achieved under the present and potentially expected operational conditions. This is similar to the familiar, but difficult, issue of QoS in communication systems. As an example, the application programmer should be able to specify that, given a particular context, for instance a life-threatening situation, a particular application should be executed with a very high degree of reliability. Another concern for the application could be the level of security or the cost. However, the PN support system oversees and guards the concurrent operation of all applications and needs to reconcile conflicting demands. Hence, there is no need for the API to expose any of this functionality. Instead, it can remain hidden from the applications.

(ii) The application programmer should have access to context information to use in the application program. This is where detailed information may be required to let the application logic take full advantage of the context. The other direction is equally important: the API should allow applications to feed context information to the PN, to be used by other applications and by the PN middleware to optimize the overall PN operation. For example, an application might need information provided by a sensor network, for example, a body or a home sensor network, to control a health application or a climate control system. Detailed information that provides context for the way quality of an application can be achieved should not be exposed, because of the risk of detrimentally affecting the overall PN functioning. This only enables the making of resource allocation decisions in isolation without taking into account the concerns of other concurrent applications. The latter responsibility lies with the PN support system and should be exposed minimally through the API to the applications.

(iii) By nature, a PN is very dynamic. Again, we feel that dealing with the dynamics should not be the task of the application programmer. However, some information on the dynamics needs to be exposed, such as the fact that essential resources for a particular application become unavailable due to mobility. An example could be a home video surveillance application that can be used remotely. When the connectivity with the user gets lost, it is useful if the application is informed and able to take particular measures to cope with the situation.

(iv) The services and resources available in a PN should be made visible to its applications. However, because of the potential redundancy and the dynamics of these resources, such as their changing availability due to device mobility, the application programmer should not be aware of the potentially very complex system state associated with a particular resource. Rather, virtual resources should be defined that, depending on context, the state of the PN, and the quality requirements of the application, are mapped onto real resources by the PN support system. For example, an application may want a multi-modal display as an interface to the user. The application could be written for a virtual display and require the best possible quality, given the circumstances. This could result in a mapping onto a handheld device display when the user is out walking, a car stereo system when the user is driving, or a home entertainment system when he is at home.

It should be clear that the design of an appropriate PN API is the subject of ongoing research and development, with several unknowns still to be addressed. Nevertheless, in the rest of this chapter, we will highlight current efforts in this direction as well as potential solutions.

8.2.2 Mechanisms for Supporting the Applications

As explained earlier in this chapter, several mechanisms are needed to provide the required functionalities to the applications running on a PN. Some of these mechanisms are exposed

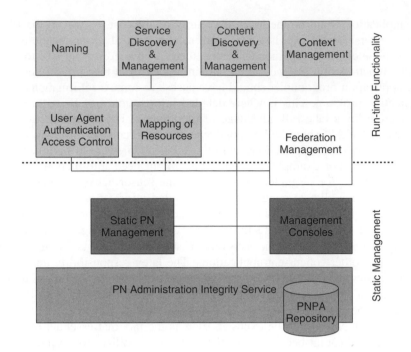

Figure 8.2 Functional decomposition of the PN support system (PNP2008 2008e).

through APIs while others should be hidden. A generic functional architecture for application and service support was proposed by the PNP2008 project. It consists of a number of components similar to what is shown in Figure 8.2. We have extended this architecture to include more components and restructured some of them for improved cohesion. Nevertheless, the functional decomposition in Figure 8.2 is still based on the existence of an administration for which the management information is guaranteed to be consistent. In the remainder of this section, we will discuss these components in more detail and what we expect them to do.

8.2.3 Naming

The naming component implements name resolution for PN services and applications. It allows the user to define names for personal devices, services, PNs, etc. and use the names from any application.

8.2.4 Service Discovery and Management

The purpose of the service discovery and management component is to help applications find suitable services. This allows applications to be decoupled from hardware functionalities, such as a microphone or a monitor, which can offer their functionalities as services. It also makes applications more flexible as they may be dynamically able to switch from one service to another based on the dynamics of the user and his PN.

Service discovery is intended to provide a secure environment, where the personal nodes that form the PN can advertise the resources and services they offer to PN applications and the PN user, and PN nodes and applications can discover which services are available at the moment. The search for services may in general yield multiple results. It may be up to the user to make the choices that are most appropriate, or an alternative is to have the PN take care of making some of the choices, using context information in the selection process.

The responsibility of service management is to manage service sessions so that conflicting demands are avoided. It may terminate running service sessions by telling the application to stop using a particular service. It may also inform the application about other alternative services that can be used. If the service management functionality is context-aware, it allows for an application to switch between monitor services based on user movements.

In Section 8.3, we will discuss in more detail the implementation of service discovery and management.

8.2.5 Content Discovery and Management

Given the wide scope of the applications expected to run on PNs, one has to deal with a great variety of types of content, with different characteristics in terms of importance, required storage, sensitivity, the time dependency of its value, etc. Examples are personal data ranging from personal music or video files for entertainment, to real-time medical data. Application requirements regarding their access to content may also vary widely. Access to content may have to be real-time, for instance in a remote health monitoring application, or may be delay tolerant, as when downloading a fresh supply of music on a car entertainment system prior to going on an trip.

Another perspective on content is the heterogeneity and the physical location of the personal devices that are the origin of a particular piece of content and those that use or store this content. The personal devices that store content can also have vastly different capacity and speed. Add to this the dynamics of a PN and it is obvious that a content discovery and management facility should be part of the application and service support in a PN, in order to facilitate the development and running of PN applications, in which content plays a big part.

A PN also offers opportunities, in terms of storing and delivering content, that are not available in non-PN personal environments. One can treat a PN as if it were a grid-computing environment (Buyya and Venugopal 2005) for personal content. The multitude of personal devices with storage capacity allows for optimizing the storage of content, for instance reducing access time, increasing availability, maximizing delivery speed or protecting privacy-sensitive data. Let us take a closer look at this.

An application might need to deliver content within a short delay to a personal device held by a mobile user. This could be achieved by the PN caching that content dynamically to a capable personal device that is always close to where the content will be used. It might even do this in a proactive way by predicting where the user is going to be in relation to the location and connectivity of his personal devices. The content should be replicated such that it is always accessible where it is needed.

Further, when delivery speed is important for large content files, one might consider distributing and replicating the content over several personal devices, as is done in peer-to-peer networks, to increase the delivery bandwidth.

Some content can be very important. In this case, the PN should store this data in a personal device that is likely to be safe from a physical and information security point of view. Unless, the content is very large, keeping automatic backups as well as older versions of the file on several personal devices will minimize the likelihood of information loss, also when personal devices fail or are lost.

From the cases described above, it is obvious that content management needs the support of other PN application support components. For instance, it needs to be context-aware and it can profit from virtualization of PN resources with the optimized mapping of virtual resources onto real resources. These functionalities are discussed below.

8.2.6 Context Management

This component enables services and applications to exploit context information. The specific task of the context management component is to collect this data and make it available in an easy-to-use way for PN applications, services, and other PN mechanisms. It needs to provide a secure environment where personal nodes make context information available to other personal nodes and to their applications. Context management includes gathering, modification, processing, storing, and propagation of context information.

The implementation of a context management system will be further discussed in Section 8.4.

8.2.7 Mapping of Resources

Building on functionalities, such as service, content, and context discovery and management, we can enable the virtualization of PN resources. We can use PN optimization and cognition to deal with the PN dynamics. A view which puts these mechanisms into perspective is shown in Figure 8.3.

As described before, the PN environment opens up the potential of choice among redundant resources with equivalent functionality, but not necessarily the same quality. PNs are inherently cooperative environments, since the personal devices all belong to the same owner. This should be helpful in achieving PN-wide optimization on behalf of the complete set of applications running on the PN.

The application specifies which resources are needed and the quality desired, subject to the context. The PN application support system should have the mechanisms to find the physical resources that are best suited given the quality demands, the context and taking into account the global view of all applications and their potentially competing and conflicting demands. This is the task of the mapping component in Figure 8.3. It should be able to continuously remap virtual resources onto the best real resources to keep up with the PN dynamics. Again a global perspective across applications is needed and interaction with the application itself may be necessary.

This is a functionality that may greatly help to improve the perceived quality of the applications. It should be able to anticipate changes in resources and context, and either

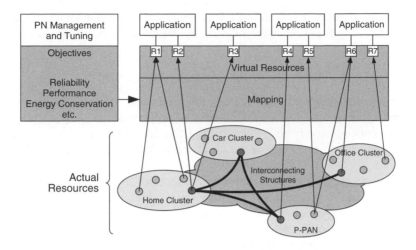

Figure 8.3   Virtualization of PN resources and their mappings.

deal with them in a timely fashion or exploit the knowledge about what is to be expected to plan ahead and optimize the system performance. This will also help to deal with complex and evolving or unforeseen situations, which are hard to cope with in an algorithmic way by, for instance, recognizing trends or previously encountered critical situations. Figure 8.3 shows four applications using seven virtual resources. The mapping functionality maps the virtual resources onto the real resources on the personal devices as shown in the figure by the lines. The mapping, which may be dynamic, is decided by objectives and preferences that are determined by an intelligent optimization management function, by the user through a user interface, or a combination thereof. It uses collected context information to make its decisions and to keep up to date so that continuous remapping can be done when needed.

> *Every now and then, Jane enjoys listening to music, especially when she is on the move. One day, when she is going home after a long day's work, she decides to turn on her music player application. It accesses Jane's favorite music from her media server at home and matches the available songs with her taste and current mood. Then, the application tries to find a suitable loudspeaker system for playback. For this, it uses a virtual loudspeaker resource. As Jane first puts on her headset, her PN assigns the virtual loudspeaker onto her headset. As she reaches her car and takes her headset off, the PN reassigns the virtual loudspeaker onto the car sound system. Finally, when she arrives home, it is mapped onto her home hi-fi system. Note that the music player application is simple and the advanced features of being context-aware and switching between loudspeaker resources are moved to the PN application support system.*

It should be remarked that the optimization of the mapping of virtual resources onto physical resources as seen by the application, will in general be a responsibility shared between the application itself and the PN application support system. The reason is that

the PN application support system, which is generic, should not know the application
logic and therefore will not be aware of certain conflicts and tradeoffs that might exist.

8.2.8 User Agent, Authentication, and Access Control

The user agent manages sessions between PN users and PN entities, such as nodes,
services, and applications. Besides taking care of authentication, in particular verifying
that the user is not somebody else, PN-specific tasks are performed, such as creating and
maintaining applications on behalf of the user, and making sure they run in the right user
session.

The task of access control is to ensure that services and content files can only be
used from within a PN user or application sessions with sufficient authorization. MAG-
NET implemented access control for services, which could be decided based on policies
and a policy engine (Zeiss et al. 2007). The policy engine was based on Euler N3
(http://eulersharp.sourceforge.net/) and was used for several things where inferring based
on rules and policies were needed. Access control was one of those things.

PNP2008 also proposed a fine-grain access control (PNP2008 2008c) that is policy-
based, or rule-based as it was called in the project. A service request is intercepted by
a policy enforcement point, which sends the request together with context information
to the policy decision point, which evaluates the request taking into account the context
information and eventual constraints imposed by the PN administration for the service
that is requested.

The proposed solution allows parties, which have different business roles – owner of
the PN, provider of services, etc. – to express and impose their constraints. It also allows
for the use of different languages to express the constraints, such as XACML (Moses
2005) or X-RBAC (Joshi 2004). In this way, we can capitalize on research that is being
done in this area.

8.2.9 PN Federation Management

This component implements the PNF-related functionalities, such as the federation
agent and the federation manager. These and other PNF-related topics are described in
Chapter 10.

8.2.10 Static Management of PNs

The purpose of the management components is to optimize the availability and usability
of services and content within PNs. Several methods have been proposed to achieve this
by the PNP2008 project (PNP2008 2008b), but there are no implementations yet. Three
architectural components were defined:

Management Consoles was created to support the different business roles surrounding
a PN that are envisaged, such as user and service provider. It plays a similar role to

those played by the User Agent, Access Control, and Authentication components for the PN user, but it does it on behalf of other stakeholders, such as service providers.

PN Provisioning Administration (PNPA) is an administration that registers the services that are offered by the personal devices. Typical information it might have is: information regarding configuration data, statistics with respect to service usage, availability of services, references to equivalent services that provide the same functionality. It might also provide functionality supporting the selection of services, to the user or to embedded applications. It should be able to use context information to guide this process.

PN Administration Integrity Service is needed in order to ensure the integrity of the PNPA. This is crucial since the PNPA is a distributed database spread over many personal devices.

8.3 Service Discovery and Management Implementation

Let us now look in more detail at service discovery and management and context management, since these have been studied in detail in the MAGNET and PNP2008 research projects and have also led to concrete proposals and prototype implementations. In this section we will discuss service discovery and management, and in the next section context management.

Both MAGNET and PNP2008 studied service discovery. However, MAGNET focused much more on this topic. Not only did MAGNET propose architectures and protocols for PN service discovery, but it implemented many of its proposals in a PN prototype. In this section, we will discuss the MAGNET and PNP2008 proposals.

A more detailed description of the MAGNET service discovery and management architecture and protocols is available in MAGNET Beyond (2008b). For PNP2008, see PNP2008 (2008c,f).

8.3.1 Service Tiers

Both MAGNET and PNP2008 have layered the service discovery into several tiers or scopes. MAGNET proposed a five tier approach, while PNP2008 worked with a three scope approach. Table 8.1 describes the different tiers and how they relate to the three scopes in PNP2008. Figure 8.4 shows how the MAGNET tiers relate to the PN architecture. Each of the different tiers has different characteristics and sometimes uses different service discovery techniques.

Figure 8.5 shows the architecture for service discovery and management for MAGNET. The proposal for PNP2008 is very similar. The MAGNET service discovery and management is a hierarchical architecture, where service management nodes (SMNs) are central authorities in the clusters for service discovery and management. They are responsible for the personal services within the cluster (Tier 1 and 2) as well as available foreign services directly connected to the cluster (Tier 3). Between clusters, a peer-to-peer scheme is used. The SMNs form a peer-to-peer overlay network, and communicate as equal peers.

Table 8.1 The five service discovery tiers and their corresponding PNP2008 scopes.

MAGNET Tier	PNP2008 Scope	Network	Comment
1	Cluster	Single Communication Domain	Some link layer technologies, for example, Bluetooth, implement their own service discovery. This tier captures this fact and allows each communication domain to use their own native service discovery mechanism.
2	Cluster	Cluster	A cluster may consist of several communication domains, since we allow heterogeneity in the link layer. This tier combines the tier 1 service discovery mechanisms, which may use different technologies, into one single mechanism for an entire cluster. Services discovered in this tier are all services on personal devices.
3	–	Immediate Environment	It must be possible to find and use services on foreign nodes in the immediate environment. Tier 3 service discovery will facilitate this requirement.
4	PN	PN	To discover services in the PN, but not in the cluster, PN-wide mechanisms must be used. This tier covers the service discovery in all connected clusters of the PN.
5	World	Global	We may also want to use services that are outside the PN, but not in the close vicinity. This tier handles the service discovery on a global scale by utilizing infrastructure-based service discovery mechanisms.

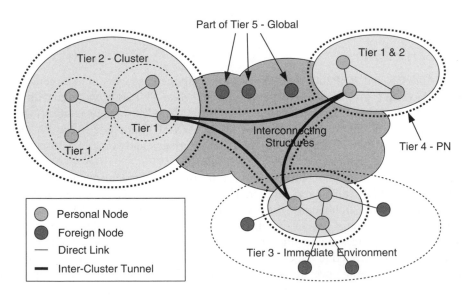

Figure 8.4 Multi-tier discovery approach.

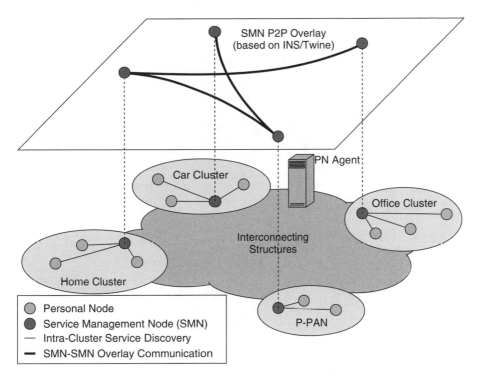

Figure 8.5 PN-wide service discovery in PNs (MAGNET Beyond 2008b).

Together with the PN agent, they are together all responsible for all services within the whole PN (Tier 4). For foreign communication, the PN agent can interact with the SMNs to locate remote foreign services (Tier 5).

Each cluster should have an SMN, which is elected among the SMN-capable nodes in the cluster. The SMN acts as a service repository and provides both service discovery and session management for the entire cluster. At the PN level, it interacts with other SMNs in the other clusters in a peer-to-peer fashion as shown in Figure 8.5. Note that the SMNs in the different clusters exchange service descriptions so that services can be discovered across the whole PN. The peer-to-peer communication between SMNs was based on the INS/Twine framework (Balazinska et al. 2002), which is based on the Intentional Naming System (INS).

8.3.2 Service Discovery Architecture

In MAGNET, the service discovery and management functionality was defined in an architectural entity called MAGNET Service Management Platform (MSMP). MSMP provides a secure environment where, on the one hand, the personal nodes that form the PN can advertise their resources and services to the PN applications and to the PN user, and, on the other hand, it provides the means to manage those resources and services. The functional architecture of MSMP is depicted in Figure 8.6.

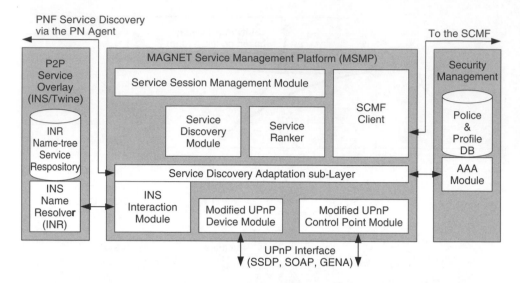

Figure 8.6 The MAGNET Service Management Platform (MAGNET Beyond 2008b).

Let us briefly characterize the role of each MSMP module or module type in Figure 8.6:

INS Interaction Module interfaces with the peer-to-peer service overlay, which creates a
distributed service repository across the different clusters forming the PN. The overlay
provides name resolution and device and service localization functionalities.

Legacy Service Discovery Modules are introduced to handle the communication between
the MSMP and legacy service discovery and management frameworks, for example,
Bluetooth service discovery and Universal Plug and Play (UPnP) (ISO 2008). The two
UPnP modules at the bottom of Figure 8.6 are examples of legacy service discovery
modules.

Service Discovery Adaptation Layer acts as an adaptation layer between the legacy
service discovery modules (e.g. the Modified UPnP Device Module and the Modified
UPnP Control Point Module) and the PN components that provide or consume services
on the one hand, and the Service Repository, upper layer components and Security
Management on the other hand.

Service Discovery Module is responsible for finding the services, that is, returning to
the application, the description of all attributes needed for an application to locate the
service and to interact with it.

Service Ranker provides the user with a ranking of the applicable services, based on
context information. For instance, if an application needs a printing service, the location
of the user can be used to find the nearest printer that satisfies the requirements.

SCMF Client provides the interface with the Service Context Management Framework
(SCMF) described in Section 8.4. It is used for passing related context information to
the MSMP modules, such as the Service Ranker for achieving context-aware ranking.

Service Session Management Module implements functions related to the management of ongoing service sessions. We discuss this module more in detail in the next
subsection.

8.3.3   Service Session Management

The Service Session Management Module maintains a list over all ongoing service sessions between clients and services in order to monitor and control them. The main reason for doing service session management is to make sure that all service sessions operate satisfactory. When too many service sessions have to share the same resources, it can be beneficial to terminate some or divert some sessions to alternative services.

This functionality is centralized at the SMNs, where all services are known. By modifying the service discovery messages, all service-related control messages can be diverted to always pass through the SMN. Because of this, the SMN will know about all clients and their ongoing service sessions. Based on context information or other knowledge available at the SMN, certain actions can be taken by the SMN to control the service sessions, including terminating an ongoing service session or diverting a client to another service.

8.4   An Implementation of Context Management

Context, together with user profile information, is an important influencing factor for the behavior of PN functions and in particular those related to applications and services, was a major concern in the MAGNET project. Therefore a generic Secure Context Management Framework was extensively researched, performance studies were done, and scalable solutions were implemented and tested in an integrated MAGNET prototype, using real applications. Details of this research can be found in MAGNET (2005a), Olsen (2008), and Sanchez et al. (2006).

Let us briefly discuss the main features of the SCMF. First, a distributed architecture was proposed, which matches the cluster-based architecture of the PN. Like MSMP, it consists of an elected context management node (CMN) in each cluster. The CMNs interact on a peer-to-peer basis with each other to provide PN-wide context services to applications. The whole framework is self-organizing and adapts to the dynamics of the PN. Secondly, SCMF was designed to be modular so that new sensors and other context sources easily could be added and in which applications can easily obtain the context information collected. The task of the SCMF is to store, process, and deliver context information from the context sources to the context-aware applications in the PN.

In addition to the CMNs, each personal node has a context agent. The context agent is an architectural entity that interacts with the applications and receives and answers their queries among others. An XML-based (W3C 2008) context access language was defined for this purpose.

The context agent interfaces with the context sources, such as sensors, operating system status of the device, physical or MAC layer parameters, and other data sources. It takes care of security, and processes and stores context information. It processes context requests and communicates with other nodes to gather the context information that is not available and integrates this information to answer the queries.

Moreover, SCMF takes into account the heterogeneity of the personal nodes in a PN by defining two versions of context agents: basic context node and enhanced context node. The latter has the full capabilities defined for context agents, while the former is for personal nodes with limited processing and storage capabilities.

For more information about the MAGNET implementation of SCMF, see Section 11.3.5.

8.5 Summary

In this chapter, we have discussed PN application support. A PN is complex and very dynamic, and that makes it challenging for the application developer to exploit its full potential. To ease the writing of PN applications, it is necessary to introduce a rich PN application support system that can shoulder many of the burdens of the applications, such as dealing with dynamics, context management, using cognition to respond to context changes, and intelligent service and content management.

We looked at a number of mechanisms that can help the applications and make them simpler and more coherent. We discussed naming, intelligent management of content, such as documents and multimedia files, and continuous mapping and remapping of virtual resources onto the most suitable physical resources. In particular, we studied in more detail the solutions proposed or prototyped by the MAGNET and PNP2008 research projects for service discovery and context management.

9

Personal Network Security

Security and privacy are essential for the acceptance of PN technology. The security architecture of PNs needs to provide security services to protect the integrity and confidentiality of data, authenticate devices and users, authorize access to services and resources, and ensure a good availability of the services. In this chapter, we discuss the security architecture at PN level.

This chapter is organized as follows. Section 9.1 discusses the provision of security credentials to a new device, such that it is able to become a legitimate member of the PN. Section 9.2 discusses the establishment of secure communication links for unicast and broadcast traffic within the PN, while Section 9.3 addresses security in the context of foreign communication. Section 9.4 deals with privacy and anonymity in PNs. Finally, Section 9.5 summarizes the chapter.

9.1 Device Personalization

The concept of the personal node is fundamental in PNs. The PN security architecture needs not only to use security services to protect the integrity and confidentiality of data, but also mechanisms to distinguish personal nodes from other nodes to be able to confine the PN to those nodes that have been 'personalized' by the user.

> Jane's sports watch has a Bluetooth interface that allows the registered data to be uploaded to a computer where various training tools can make use of it. Jane wants to incorporate the watch into her PN such that it can continuously send data to her PC via her Bluetooth enabled mobile phone, while she is training. For this purpose, she needs to 'personalize the device', that is, make it a legitimate member of her PN. After having purchased the watch, Jane removes it from the factory packaging and places it in the proximity of her phone. The telephone displays a four-digit number that Jane enters into the watch. After a few seconds, the telephone indicates with an audio signal that the watch has been successfully incorporated into her PN.

The process of associating a device with the personal network is called *device personalization*. The association is based on cryptographic material that is obtained during the

Personal Networks: Wireless Networking for Personal Devices   Martin Jacobsson, Ignas Niemegeers and Sonia Heemstra de Groot
© 2010 John Wiley & Sons, Ltd

first phase of the configuration process, called *imprinting*. After imprinting, the device can perform mutual authentication and establish a secure channel with other nodes in the PN, which in turn can be used to perform further configurations.

9.1.1 Imprinting

Imprinting is based on the resurrecting duckling security policy model of Stajano and Anderson (1999) and Stajano (2000). A device is either in the imprinted or unimprinted state. A new device (the duckling) starts in the unimprinted state where it waits until the PN user provides it with cryptographic material (coming from the master device, that is, the mother duck). After that, the duckling can authenticate commands coming from other devices of the user and reject other commands. Once imprinted, the device rejects further attempts to imprint it. The first device that offers imprinting services to the new device will be accepted as the mother, hence the terminology. When the user wants to evict a device, she sends an authenticated kill command to remove the cryptographic material and other personal information from the duckling and returns it to the unimprinted state.

> *Jane's company has decided to upgrade the employees' laptops. This means that Jane needs to hand in her old laptop. However, before she does so, she needs to remove it from her PN so that no one can use her old laptop to access her PN. To do this, Jane uses her telephone. She launches a special PN management application on the phone, selects the eviction functionality, and selects her old laptop. The phone asks Jane to prove her identity, which she does using voice recognition. Then the phone again asks whether Jane really wants to remove the laptop from the PN, to which Jane answers yes. The phone then informs all the nodes in her PN about the change. When the laptop itself receives the change notification from the phone, it again enters the imprintable state, in which the next user may add it to his or her PN.*

The general approach to imprinting is to use the principles of public key infrastructure (PKI) adapted to the PN environment. Instead of using global certificates issued by a third party, the certificates are signed by the mother device that acts as the PN certification authority (PNCA). During imprinting, the PNCA signs the public key generated by the duckling and stores the generated certificate and the PNCA certificate in the duckling. After this step, all the nodes in the PN can authenticate each other by checking each other's certificates.

PKI lightweight security solutions using elliptic curve cryptography (ECC) have been proposed in MAGNET Beyond (2006b). ECC provides the same level of security as other public key algorithms but with smaller key sizes, making it more attractive for those PN devices that are limited in resources. The MAGNET imprinting procedure using public key cryptography based on ECC as described in more detail in Section 9.1.3.

An alternative approach to public key cryptography is to provide the duckling with a shared secret with the mother. The secret can be used as a pairwise symmetric cryptographic key. The drawback of this approach is that it requires pairwise keys between all the nodes in a PN, which strongly limits its scalability.

The security of a PN depends on the security of the imprinting procedure, which is based on the assumption that it is under full control of the user. The user of the PN determines when a new device is personalized and included as member of the PN.

9.1.2 Imprinting Using Location Limited Channels

Imprinting needs a form of authenticated Diffie–Hellman key agreement protocol to avoid man-in-the-middle attacks, that is, to ensure that the security associations are set up between the legitimate peers and not with an attacker.

A way to do this is by using a location limited channel (LLC) as originally proposed in Stajano and Anderson (1999) and Balfanz et al. (2002). An LLC is a communication channel that provides special security properties such as confidentiality, integrity, and authenticity because of its physical characteristics.

The properties required for an LLC when used for imprinting are discussed in Balfanz et al. (2002). First, the channel must support identification based on the physical context (e.g. the PDA in front of me, the sensors in this room). Communication technologies that have physical limitations, such as limited transmission range or directional characteristics, are good candidates. A second property is channel authenticity, which means it is virtually impossible for an attacker to transmit in the channel without being detected by the legitimate participants. Observe that secrecy is not a requirement. Examples of LLCs are:

Wired connection. A wired connection between the mother device and the device to be imprinted, under the assumption that no physical tampering takes place, provides not only authenticity but also confidentiality.

Physical contact. Touching one device with another causes no ambiguity as to which entities are involved.

Audio. Audio, both in the audible and ultrasonic range, has limited transmission range and can be confined within a specific location.

Infrared. Infrared signals have limited transmission range, are directional, and cannot penetrate walls.

Near field communication (NFC). This form of communication makes use of the magnetic field induction of two antennas located in the near field of each other. NFC has a range shorter than 20 cm. This is a technology mainly aimed at payment applications for mobile phones and is currently being integrated into several handsets for the mass market.

User reading from one interface and entering in another. If one of the two devices has a display and another a keyboard, the user may be asked to read information from the display on one of the devices and enter it on the keyboard of the other device.

The exchange of messages does not need to be real-time. One could conceive of situations (MAGNET Beyond 2008a) in which a new low-end device without a user interface is delivered to the user with a hash of its public key stored in an external medium, such as a piece of paper or in a RFID tag. The user, acting as an LLC, later enters the hash into the mother device.

Observe that many of the above channels are severely constrained in their capacity. Reading an alphanumeric number from one device and entering it into to another device imposes a limit on the length of the string that can be transferred by the user. If a low-rate authenticated channel is used, the full-key agreement protocol can use an additional non-authenticated channel. Several protocols have been proposed that use the combination of a non-authenticated channel and a low-rate LLC (Balfanz et al. 2002; Mirzadeh et al. 2008b; Vaudenay 2005). The LLC is strictly necessary for authentication purposes in the key exchange phase. The rest of the key exchange procedure, which involves the exchange of large public keys, takes place via the non-authenticated channel.

9.1.3   Certified PN Formation Protocol

One of the key agreement protocols especially designed for personal networks is the Certified PN Formation Protocol (CPFP) (MAGNET Beyond 2006b) that was developed in the IST MAGNET BEYOND project. It consists of two stages. The first stage is the imprinting stage. The second stage is the generation of the pairwise keys between PN nodes.

The basic idea of the CPFP imprinting procedure is that the public keys are exchanged between the PNCA and the new device using a non-authenticated wireless channel. Subsequently, the keys are authenticated using an LLC. To describe this, we use the notation introduced in Table 9.1.

A distinction is made between two types of LLCs, private and public. A public LLC satisfies the requirements of authenticity discussed in Section 9.1.2. A private LLC additionally provides confidentiality. Based on this distinction, two versions of the CPFP imprinting stage have been defined:

Imprinting over private LLC. In this version of imprinting, after the PNCA and the device have exchanged public keys over a non-authenticated channel, the PNCA generates a random key K and uses it to compute a message authentication code (MAC) of the exchanged public keys. Figure 9.1 shows this version of the protocol for the case where the PNCA device has a display and the imprinted device a keypad. The user transfers the key K and MAC to the device being imprinted. The key K is used by the device to compute the MAC and compare the result with the value entered by the user. By means of an audio or visual signal, the device indicates whether the imprinting was successful or not.

Table 9.1   Notation used in this chapter.

SK_{PNCA}	Secret key of the PNCA
PK_{PNCA}	Public key of the PNCA
$CERT(PK_{PNCA}, PNCA)$	Certificate on PK_{PNCA} signed by PNCA
SK_A	Secret key of device A
PK_A	Public key of device A
$CERT(PK_A, PNCA)$	Certificate on PK_A signed by PNCA
$MAC(K, M)$	Message authentication code for message M using key K
$hash(M)$	A secure hash of message M
$M_1 || M_2$	Concatenation of the messages M_1 and M_2

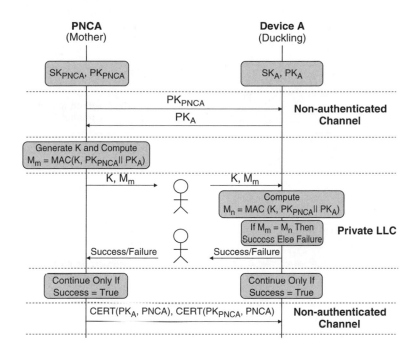

Figure 9.1 Imprinting over a private LLC.

Imprinting over public LLC. This version of imprinting is more generic since it does not impose extra requirements on the LLC. After exchanging the public keys over a non-authenticated channel, the PNCA generates a hash of the public keys and sends it via the LLC to the device being imprinted. The device also computes the hash of the public keys and compares it with the received value. The result of the procedure is given to the user via a signal. Figure 9.2 shows this version of imprinting.

In the second stage of CPFP, the PN nodes use their certificates to authenticate each other and establish *master pairwise keys*. A master pairwise key is a shared key between two personal nodes in a PN. It is used to generate session keys to be used when the two nodes communicate with each other, which we will discuss in Section 9.2. To establish the master pairwise keys, any standard key agreement protocol can be used.

9.1.4 Eviction of Personal Nodes

The certificates issued by the PNCA have a limited lifetime after which they are no longer valid and need to be renewed. Sometimes, it is necessary to evict a personal node from the PN before the lifetime of its certificate has expired. There may be various reasons for evicting a node. Examples include device loss, returning a rented device, a change to the device name or suspected compromise of the private key.

The way to evict a node is to revoke its certificate. The standard method for making the information on revoked certificates known is by issuing a certificate revocation list (CRL)

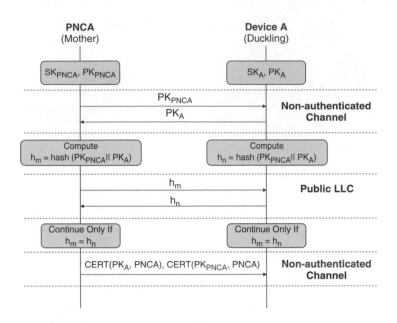

Figure 9.2   Imprinting over public LLC (PNP2008 2008a).

and distributing it to all the nodes in the PN. In a PN, it is the user herself who decides whether a node has to be revoked or not. The CRL is signed with the private key of the PNCA, which allows other PN nodes to check the authenticity of the CRL. PN-wide broadcasting, as described in Section 6.6.3, can be used to distribute the CRL in the PN.

9.2   Establishment of Secure Communication

When two personal nodes meet and are able to communicate with each other, it is important that they first authenticate themselves. This can be done by using the master pairwise key if they share one. This is a more efficient method since it only involves symmetric cryptography, which in general is less computationally heavy. Should there not yet be an established master pairwise key between the two personal nodes, then they can fall back on the certificates. Using the certificates, they authenticate each other and can agree on a master pairwise key. This will also take place if the master pairwise key is expired.

When a neighbor node has successfully been authenticated as a personal node, it is necessary to establish a secure communication link for both unicast and broadcast traffic. The master pairwise key is only used to derive or securely exchange a link layer session key for data encryption and integrity protection. How the link layer session key is established depends on the link layer technology.

9.2.1   Secure Unicast Communication

The various link layer technologies should use the session keys to encrypt and decrypt the packets at the link layer whenever the link layer provides adequate security mechanisms.

Hardware-based encryption provided by the network interface cards can then be used, and this improves both performance and power consumption. For an end-to-end path between two personal nodes within the same cluster, this encryption is performed on every hop. This is the way intra-cluster unicast traffic is protected. This method is fully ad hoc and fully distributed as well as secure if adequate encryption is used on each hop.

In the case of Bluetooth, the length of the key is defined to be 128 bits, while in IEEE 802.15.2, it is 256 bits. A session key derivation generation procedure based on the master pairwise key is executed. This procedure typically results in pairwise session keys. An example is the four-way handshake protocol of IEEE 802.11i (IEEE 2004a). If there is no data encryption and integrity protection provided by the link layer, this has to be implemented at the network layer instead.

Further developments of these mechanisms by IST MAGNET are reported in MAGNET (2005c).

9.2.2   PN Awareness at the Connectivity Level

PN and cluster formation must work on link layer technologies that are unaware of PN boundaries. At the same time, many link layer technologies form piconets or other types of logical structures. They do that to better mitigate collisions and contention on a shared wireless channel or to enable power saving modes. However, this may lead to some unwanted situations. Consider the two scenarios shown in Figure 9.3. The figure shows the clusters of two different persons, which means that they must never merge. The first scenario is what may happen if Bluetooth is used as one of the link layer technologies (other link layer technologies may produce a similar result). The Bluetooth link layer is unaware of the PN boundaries and may easily form a single piconet consisting of nodes from both PNs. A piconet controller may control the medium access of nodes of another cluster.

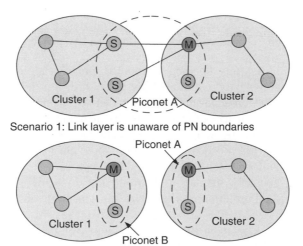

Scenario 1: Link layer is unaware of PN boundaries

Scenario 2: Link layer is aware of PN boundaries

Figure 9.3   Link layer aware or non-aware of PNs.

Future link layer technologies may of course be specifically designed for PNs and hence understand PN boundaries and cluster formation requirements. In the second scenario of Figure 9.3, the piconet formation protocol is aware and is forced to include only personal nodes. The second scenario provides better security, which may tempt us to suggest such a solution despite the fact that it requires changes to the link layer. However, the first scenario provides better medium access management possibilities, since it works across the cluster boundaries.

The choice whether the link layer should be PN aware is all but clear. In many cases, it is desirable to be protected against radio jamming and other types of denial-of-service (DoS) attacks. However, when the contention increases due to many devices sharing the same wireless channel, then contention and collision mitigation schemes operating between the involved clusters would be valuable. The situation can be quite bad when people gather in large crowds, such as sports events, large exhibitions, and the like. The ultimate system should coordinate medium access and at the same time protect the system against DoS attacks.

9.2.3 Secure Broadcast Communication

Not only secure unicast communication is important – secure cluster-wide broadcasting is also required. Several mechanisms require it, such as service discovery and dissemination of context information. To implement a secure cluster-wide broadcasting mechanism, we need secure broadcasting at the link layer. Link layer broadcasting is typically only one hop, but will be an important building block for cluster-wide broadcasting. In this section, we discuss how the link layer can provide this building block in a secure way. Section 5.4.4 covered the topic of cluster-wide broadcasting using flooding at the network level.

Clusters may consist of several different link layer technologies, and each of them implements its own way of securing link layer broadcasting, if they implement a secure broadcasting at all. Even if a current link layer technology is capable of secure link layer broadcasting, it is likely to be unaware of PN boundaries. Based on this, we see three possible solutions to the cluster-wide broadcasting security problem:

1. We could use a unicast-only flooding protocol (Lipman et al. 2004) for cluster-wide broadcasting and completely avoid anything to do with link layer broadcasting. This adds reliability to the broadcasting since we can use acknowledgments, RTS/CTS, etc. However, it is much less efficient in terms of number of transmitted messages, especially since clusters are compact and a broadcast medium is used.
2. Distribute a cluster-wide broadcast key in the whole cluster and use that key for all broadcast traffic. Unfortunately, distributing that key is a cluster-wide broadcasting problem in itself. The only option is to distribute the key using a unicast-only flooding technique as above. Furthermore, when two clusters with different broadcast keys merge, they have to converge to one key, which makes this proposal all the more complex. Since we depend on the security features of the link layer technologies, there are almost no benefits to this solution.
3. Each personal node has its own unique link layer broadcast key to broadcast to the immediate personal node neighbors. The broadcast keys can be exchanged with unicast messages right after two personal nodes find each other in such a way that only personal

nodes have access to the keys. For instance, broadcast keys can be distributed together when the pairwise session keys are being established.

The third option offers the best solution, since it keeps the broadcast communication to a minimum and at the same time keeps the amount of computation low. Each message needs to be decrypted at each hop in the first place, and decryption/encryption can take place in the hardware using the link layer solution. Node mobility within a cluster is also less of a problem in this solution. The broadcast key is exchanged as part of the standard neighbor discovery process and nodes may keep the broadcast key even after they have lost contact with a neighbor. If a broadcast message is received from a disappeared neighbor, it can still be decrypted, unless the broadcast key has been updated, which does not need to happen often.

9.2.4   Secure Inter-Cluster Communication

The securing of inter-cluster tunnels works similarly to the securing of intra-cluster communication mentioned earlier. In this case, the gateway nodes are responsible for providing connectivity between personal nodes of different clusters and therefore also for securing the tunnels. The inter-cluster tunnels between nodes are secured by setting up IPsec tunnels. The algorithm used for key establishment in the IPsec tunnels is similar to that used for the second stage of the CPFP: gateway nodes exchange PNCA certificates and, after successful authentication, derive session pairwise keys.

9.3   Secure Foreign Communication

Foreign communication was discussed in Chapter 7. Before communicating with a foreign node, one should first verify the authenticity of that node. This can take place as soon as the node is discovered and before being announced within the PN in order to avoid overloading the PN with information about non-trusted foreign nodes. However, the authentication operation may be quite heavy and consequently it may be better to perform it on demand instead. In either case, several authentication approaches exist, such as PKI and reputation-based approaches.

Assuming that two PNs want to communicate, both PNs could contact a certificate server (CS) to get the necessary authentication information. A CS may be a different server supported by a third party or constitute a separate functionality in the PN agents themselves. However, contacting a CS every time a foreign node needs to be authenticated is tedious and will not work when the CS is unavailable. Various caching mechanisms of the certificates can be used. The PN agent can, for instance, keep certificates of known PNs on behalf of its PN.

Another way of authenticating foreign PNs is to have it performed manually by the user when two PNs meet physically. Imagine that two persons and their PNs meet on the street. By means of proximity authenticated channels or other types of secure channels, they can physically authenticate each other. A certificate can be exchanged as part of the authentication and this can be cached just like a certificate received from a CS. In addition, these certificates can be exchanged between friends in a similar way to the web of trust as used in PGP (Zimmermann 1995).

When a foreign PN can be authenticated, it is also possible to establish a secure connection. This will make it impossible for other foreign nodes to interfere with the communication. It also includes the protection of the data itself from wiretapping and manipulation.

These are just examples of how foreign node authentication can be done. Many more options exist in the literature. For more examples and further details on these topics, see MAGNET Beyond (2006a).

9.4 Anonymity

To be able to authenticate neighboring nodes, one can rely on fixed identities. The identifier of a PN node includes both a PN identifier and a node identifier so that a receiving node knows exactly which pairwise key to use. The identifier can also be mapped onto MAC addresses or anything else that is fixed. By using a mapping table, the receiving node can know which master pairwise key to use.

However, for privacy reasons, it is not always wise to expose fixed and unique identities. Frequently transmitted identities, such as PN and node identifiers, can be linked to the user of that device and this makes the device and its user no longer anonymous. This also includes fixed network addresses. A mobile device is said to be anonymous if it does not reveal anything that can be used to link it to a person or to a previously encountered device. The latter means that it should be impossible for someone to know whether he has communicated with or overheard communication from the same device before or not.

When packets with fixed addresses are sent back and forth, it is possible for any third party to know who communicates with whom, at what time, and where. Furthermore, protocol analysis can allow one quite accurately to guess what type of data is being sent (e.g. voice, e-mail, or web) even if the traffic is encrypted. For mobile devices, the addresses may also tell something about the location of that device at the time of communication. When all this is put together from many sources, a lot of information can be deduced that may reveal a person's location, whom he is communicating with, and at what time – potentially information the person does not want to reveal. Furthermore, anyone can lay their hands on this information, not just network providers.

All this is actually no longer just a theoretical threat; a system to track Bluetooth devices was installed in September 2007 in the city of Apeldoorn in the Netherlands (http://www.bluetoothtracking.org/). Currently, the system consists of eight stations at different locations, including locations outside Apeldoorn, recording anyone passing by with a device with Bluetooth switched on. If you know someone's Bluetooth MAC address, you can find out when and if that device has been seen on any of the locations by visiting their website. Similar tools also exist for GSM phones. Hence, it is necessary to hide all fixed identities to remain anonymous.

9.4.1 Anonymity in PNs

Unfortunately, anonymity also makes it harder for the nodes to select the correct master pairwise key when encountering new neighbors. In the worst case, all stored master pairwise keys must be tested one by one before the authentication fails and the node is declared foreign. This is, of course, a waste of communication and computational resources, and better solutions are needed.

One very simple idea, which we propose due to lack of better options, is to have one PN-wide key that is shared by all nodes in a PN. The key can be exchanged during the imprinting phase so that every node in the PN can store it in its local memory. The key should only be used to encrypt PN and node identities and nothing else. The compromise of this key leads to lost anonymity, but not to access to the PN network. The latter still requires breaking the PNCA certificates or the master pairwise keys.

While it is possible to encrypt all fixed addresses and identifiers in the network layers and higher, it is also necessary to protect fixed addresses used by the lower layers. Most wireless technologies use unique identifiers that can be read and used unencrypted and it is not possible to encrypt them, since they are used to determine the sender and the receiver. WLAN (IEEE 802.11) is one example. Each WLAN card has a hardware address of six bytes that is used when communicating with another WLAN-equipped device or access point. To make sure that addresses never clash, each WLAN card is equipped with a globally unique address. This address can, of course, be used to map to the user of the device it belongs to. Furthermore, the WLAN card is constantly transmitting its hardware address unencrypted, even when encryption is used. This is still the case when using the latest WLAN encryption standards, such as Wireless Protect Access (WPA) and WPA2 (IEEE 2004a). Bluetooth suffers from the same problem, since it uses the same type of hardware addresses. However, Bluetooth is slightly better as it does not transmit its unique hardware address with every message. It is only sent as part of the piconet formation procedure after discovering a new neighboring device.

The best answer to this problem is to get rid of the uniqueness and frequently change link layer addresses, since global uniqueness is not really needed anyway. The address only needs to be unique among the devices in the near vicinity, where its transmissions can be received. Even if there are address conflicts, they can be solved by inspecting the (encrypted) network layer address or by the failure to decrypt the packets. A conflict only degrades the performance somewhat. Anyway, the addresses can easily be manipulated by a malicious user and hence cannot offer any additional security.

We should also note that just as there is no perfect security, perfect anonymity is impossible. There will always be information leakage that may be used to guess who the user of a device is. However, this must not stop us from developing techniques that make this more difficult. The most obvious vulnerabilities must be avoided which will make it much harder for an adversary to discover this kind of information. The first step is to avoid fixed addresses that are sent unencrypted and can be overheard by non-authorized peers. When that is taken care of, an attacker needs to turn to the physical layer to look for clues. One example is using a technique called radio frequency fingerprinting (Hall et al. 2003). However, such techniques are much more difficult than just listening for unencrypted fixed identifiers.

9.4.2   Anonymity in Foreign Communication

With regard to anonymity in foreign communication, things become much more difficult. The solution of having one common key for encryption of node and PN identifiers that was proposed for protecting intra-cluster communication will not work. One cannot share a key with every friend one has, without them also sharing the same key with each other. Furthermore, the solutions used for securing infrastructure-based connections will not work either. In those solutions, only one of the peers needs to remain anonymous. Hence, the infrastructure-based equipment announces its identifier and public key in the open so that clients can establish a secure link anonymously.

To allow a node to reveal its identity and establish a secure connection to another node only if there is trust between the two nodes, and remain anonymous otherwise, seems impossible if the other node also has the same requirement. Not knowing the identity of a newly discovered device, and trying to figure out if there is trust in that device, are contradictory requirements. More efficient solutions than trying each available key, one by one, are difficult to find. Possible candidates may involve the two peers gradually giving away clues to their identities, until either of them concludes that only one option remains, and tries that option. If it succeeds, there is trust, otherwise not. For the time being, this problem remains an open and challenging research issue.

9.5   Summary

Security and privacy solutions are key elements in the development and acceptance of PN technology. From the point of view of security, a PN resembles a virtual private network where only authorized devices are able to get access. However, because permanent access to a trusted third party may not be possible or desirable, traditional security mechanisms based on public key infrastructure and certificate authorities are not directly applicable to PNs.

In this chapter, we have discussed security and anonymity solutions that are tailored to the characteristics of personal networks. The solutions minimize the user interaction and take into account the limitations imposed by those devices that are limited in their resources. Ownership plays a major role since the PN is bound to those devices that have been personalized by the user.

The PN security architecture is based on bilateral trust relations between personal devices. These relations are established in two stages. In the first stage, the devices are associated with the PN by a process called imprinting. The cryptographic material obtained during this stage is used in the second stage to establish a shared secret key between PN devices. The shared secret keys are used to derive session keys in different protocol layers for both intra-cluster communication and inter-cluster tunneling.

We also touched on security in foreign communication, by outlining possible ways to authenticate foreign nodes and PNs. Finally, we discussed the concepts behind anonymity, which is a very important building block for achieving good privacy. While we were able to propose a solution for anonymity in intra-PN communication, anonymity in foreign communication still remains a challenge.

10

Personal Network Federations

As the number of resources and services that are available to the individual grows, it is likely that users will wish to make them available to others. *Personal network federations* (PNFs) extend our vision of person-oriented networking to group-oriented networking. In a PNF (Niemegeers and Heemstra de Groot 2005), users can share a selected subset of their personal resources with others for private or professional purposes under clear and well-established rules of cooperation. This chapter discusses the extensions to the PN functionality presented in the previous chapters needed to support interactive and cooperative services between multiple PNs.

A federation of personal networks, PNF, or fednet is an agreed cooperation of independent PNs with the purpose of achieving a specific common goal. The participants in the PN allow each other access to specific services and the usage of specific resources for achieving the common objective. The concept was first proposed in Niemegeers and Heemstra de Groot (2005) and was further developed in the European MAGNET Beyond (http://magnet.aau.dk/) and the Dutch Freeband PNP2008 (http://pnp2008.freeband.nl/) projects.

The resources that a PN shares in a PNF are committed only for the time and purpose of the federation. In principle, only a subset of the resources of the constituent networks are shared in the PNF. Which subset is part of the federation depends on the application and the agreed rules, for instance, what is useful or needed to achieve the common task. The concept of the PNF is illustrated in Figure 10.1, where three PNs (A, B, and C) form a single federation involving a subset of the services of each constituent PN. Without federating, each component has a set of 'initial available services'. When A, B, and C federate, each makes available a subset of their initial services to the PNF. The result is that each PN has at its disposal a larger set of services thanks to the PNF.

This chapter is organized as follows. Section 10.1 presents several examples of PNFs to illustrate their use. In Section 10.2, we provide a classification of PNFs depending on various characteristics. Section 10.3 introduces requirements for federations derived from the presented use-cases. Sections 10.4 and 10.5 respectively present the architecture and life cycle of the PNF. In Section 10.6, we discuss how access control is carried out, while in Section 10.7 we expand on the two architectural approaches for implementing PNFs. Section 10.8 deals with the security mechanisms in PNF. Finally, we summarize the chapter in Section 10.9.

Personal Networks: Wireless Networking for Personal Devices   Martin Jacobsson, Ignas Niemegeers and Sonia Heemstra de Groot
© 2010 John Wiley & Sons, Ltd

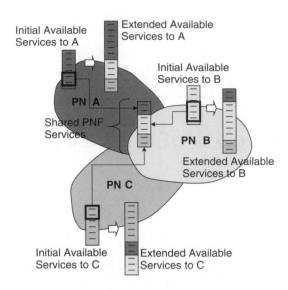

Figure 10.1 The personal network federation concept.

10.1 Examples

There are many private and professional group applications that could be enabled by this paradigm. We recall the scenario presented in Section 4.9 where Jane and three of her colleagues form a PNF after having visited a conference.

> *To strengthen their knowledge, Jane's boss decides to send Jane and three of her colleagues to a conference, as illustrated in Figure 10.2. After they return from the conference, they all want to look at and share the photos and videos that they took. For this purpose, they form a PNF that contains a subset of their PNs. This subset is composed of a camera phone and a laptop (PN1), a computer with photo editing software (PN2), a digital photo camera and a photo printer (PN3), and a digital video camera (PN4).*

In this PNF, the four colleagues can share several types of content (e.g. photos and videos) and services (e.g. display, photo-editing, and printing services) with each other. Note that because the PNs only share a subset of their resources, those PN resources that are not included in the PNF will not be visible or accessible to the other PNs. Once the four colleagues have finished sharing the videos and editing and printing the photos, they dissolve the PNF.

This is only one example. We can imagine many more in many different areas, such as education, health care, entertainment, public safety and business. Some examples are provided below:

Team of emergency workers in a disaster relief scenario. Imagine a major emergency scenario involving different professionals, such as firefighters, policemen, medical personnel, and environmental specialists. The professionals have a PN each to assist them

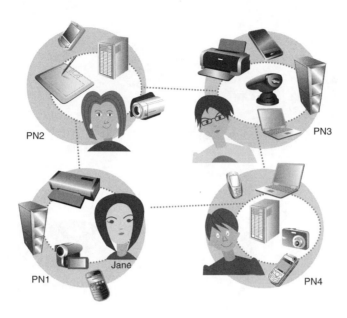

Figure 10.2   PNF for sharing photos, videos, and printing services.

in their tasks. Depending on the primary task of the professional, the PN may contain many different devices, including a variety of sensors, cameras, displays, and communication devices.

   In addition, the professionals may be geographically dispersed and have a partial view of the situation. Federating their PNs to share monitoring and observation devices will enhance the capabilities of each professional involved and help them to have more accurate information on the whole situation. For example, the policemen or medical personnel will be able to assess the situation inside a burning building by being able to access the helmet-mounted cameras of the firemen. This example is illustrated in Figure 10.3.

Sharing content and services with family and friends. Some friends meet at a restaurant and two PNFs are formed, one for sharing vacation pictures and videos among adults and a second one for the children to play computer games together. Each person commits a subset of their resources to the PNF. Some resources, such as a laptop serving as a communication hub, may be involved in both PNFs. This is shown in Figure 10.4. Note that the PNFs do not require that the people involved are present at the same location. They could be anywhere as long as their PNs have the opportunity to communicate with each other, for instance, by using the Internet.

Sharing sensor information from cars to increase road safety. Cars traveling close enough to be able to establish radio links could federate and make available their accurate position together with other sensor information to other members to increase road safety. Position-sensing devices, sensors involved in automatic braking systems (ABS), electronic stability control and traction control could provide important real-time data to surrounding cars in a PNF. This information could be used for predicting dangerous situations and warning the drivers, as in a system for cooperative forward

Figure 10.3   PNF for an emergency relief situation.

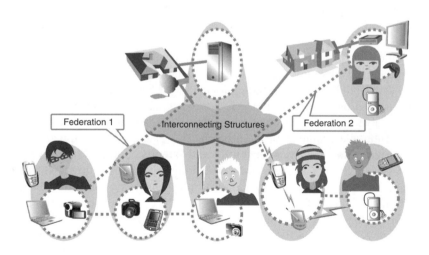

Figure 10.4   PNFs for sharing content and services with family and friends.

collision warning. Such a PNF would be a dynamic entity, where participants join and leave all the time based on mobility. This is shown in Figure 10.5.

Group applications for health care. We could conceive of a PNF composed of a person with impairments, some members of her family and friends, as well as medical professionals. The friends and professionals run different applications to support the impaired

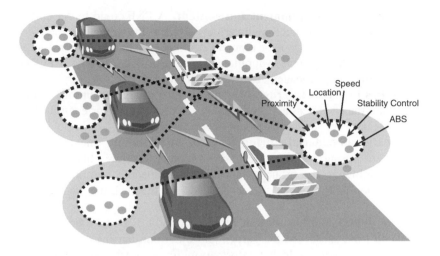

Figure 10.5   PNF for increasing road safety.

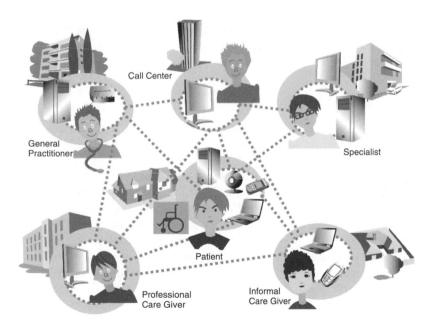

Figure 10.6   PNF for assisted living.

person and allow her to live independently. Depending on their role, the participants will have access to some resources while being denied access to others. For example, the doctor may have access to medical sensors on the patient, but not to her private photo album, whereas a friend may have access to her photos and agenda, but not to her medical data. This example is shown in Figure 10.6.

Project members sharing equipment and content. A PNF could be set up by a project
   leader to allow the various team members involved in the project to share content and
   resources, such as documents, measurement equipment, communication equipment, and
   plotters. The project may have a duration of weeks or even months. After finalization
   of the project, the project leader dissolves the PNF.

Sharing facilities to access Internet. We can imagine a large group of persons forming
   a PNF to share their wireless access facilities across a country or even further to create
   a large access network for the members of the federation.

Sharing educational material during a class. In education, one could think of an
   example where during a class, a teacher temporary federates her PN with those of
   her students, allowing the group to get access to specific training materials and to
   share information and resources. After the class has finished, the teacher may dissolve
   the PNF.

Sharing home digital weather station information. A PNF to share temperature and
   humidity data from sensors in the garden and other outdoor locations of a large group
   of people could provide valuable information for weather forecasting. It would be
   complementary to the information provided by radars and satellite, which cannot easily
   capture the situation close to the ground.

Although partial solutions for supporting some applications may exist, they usually
require substantial technical knowledge and time on the part of the user. PNFs provide a
single solution to support all applications and reduce user intervention to a minimum.

10.2   Types of Federations

PNFs can be classified according to various characteristics (Hoebeke et al. 2006a;
PNP2008 2008d). In this section, we introduce the five most important classifications.

Depending on how PNFs are initiated, they can be classified into *purpose-driven* and
opportunity-driven. In the purpose-driven case, the formation of the federation takes place
when the need to achieve a common task arises. In the opportunity-driven case, agents or
applications running in a PN are notified of the opportunity to federate and consequently
a federation is formed. The PNF for sharing photographs with friends and family is an
example of a purpose-driven federation. Sharing vehicle sensor information for increasing
road safety is an example of an opportunity-driven federation.

Based on their lifetime, a PNF may be either *short-lived* or *long-lived*. While short-lived
federations need simple and lightweight set-up and management mechanisms, long-lived
federations, in general, can afford more powerful and complex solutions. The PNF for the
emergency team is an example of a short-lived federation. Long-lived federations make it
easier to collect information on the quality of the services offered by the participants. This
information can be used to implement access control policies, service selection strategies,
and rewarding mechanisms based on the reputation of the participants. The federation for
creating a large wireless access network is an example of a long-lived federation.

Based on the way the federation process is initiated, we can distinguish between *proactive* and *reactive* PNFs. In the proactive case, the federation is established and maintained
in anticipation of a future need for cooperation in order to reduce the configuration latency
when an application is needed. A practical example would be the case where people are

having a meeting and their PNs are federated in anticipation of a group application. In the reactive case, the federation is set up when the need for cooperation arises. An example is the federation for sharing educational resources during a lesson.

Depending on whether the federation needs the support of services in the infrastructure or it is fully ad hoc, we can distinguish between *infrastructure-based* and *ad hoc* PNFs. The federation for weather forecasting is an example of the former, whereas a federation for gaming of collocated persons is an example of the latter.

Somewhat related to the previous one is the classification into *provider-enabled* and *user-enabled* federations. In provider-enabled PNFs, some of the functions needed to set up and manage the federation are carried out by one or more third parties. An example is a PNF for gaming where a third party offers a brokerage service by matching interest of potential participants to initiate federations for playing a specific game. The broker itself does not participate in the PNF. In the user-enabled case, the PNF is formed on the initiative of one of the users as in the case of a federation for sharing photographs.

10.3   Requirements

Based on the examples of the previous sections, it is possible to derive a set of requirements for PNFs (Hoebeke et al. 2006a) as follows:

Policies for controlling the access to the PNF. Group cooperation requires the management of the access policies for participating in the federation. In addition, access to services and resources should be limited to the participants in the federation. This means that the federation should have an entity that takes care of performing access control by authenticating PNs that want to participate in the federation and consequently applying the corresponding authorization policies.

Mechanism for automatic setup, organization, and maintenance. For reasons of usability, the setup, organization, and management of PNFs should be a process that is hidden as much as possible from the users. This requires the definition of polices and rules that indicate under what conditions a PNF can take place and mechanisms to make this available to potential participants.

Management of the membership and resources. The composition of a PNF is dynamic and so is the availability of the services and resources provided by the participating PNs. An entity should keep track of the current set of participants and the availability of the resources they have committed to the PNF.

Identity management. Independently of the authentication method used to provide access to new members, there should be a way to identify the PNs and their resources within the local scope of the PNF.

Service discovery within the scope of the federation. The participants in the PNF need to be able to discover and localize the services and resources they need.

Security and privacy. In addition to measures to control the access to the PNF and depending on the application, additional security services may be required, such as confidentiality and data origin authentication. Furthermore, it should not be possible for a participant to access services and resources or even discover them if they have not been committed to the federation. In some situations, it may be important not to disclose the internal network structure of a PN to the other participants in the federation.

There are also some extra requirements concerning the PNs that participate in a PNF:

Mechanisms for joining and participating in PNFs. Complementary to the mechanisms for self-organization and maintenance at the PNF level, mechanisms at the PN level are also required, such as rules and policies to determine how and when a PN can and should participate in a PNF.

Management of the resources committed to the PN. Mechanism are necessary to specify and manage the resources and services that are made available to federations. Observe that it should be possible for a PN to participate in several PNFs. The establishment of a federation should preferably be a process that does not require user involvement; however, for human trust reasons, user involvement may be desirable in certain cases, for instance, by notifying the user or asking for formal approval.

In addition to the functional requirements, there is also at least one performance requirement:

Scalability. The PNF concept enables many different potential scenarios. Some of them may involve a huge number of participants. From the point of view of the PN, it may participate in many different federations simultaneously. Therefore, solutions are needed that provide good scalability, meaning that the user experience is not severely affected when the number of federations in which a PN participates and the number of members that a federation supports increase.

10.4   Architecture of a Federation

A PNF is a secure cooperation between different PNs in which selected services and resources are shared. The architectural components that are needed to support the requirements stated in Section 10.3 are shown in Figure 10.7 for the case where a PNF is managed by one of its members and Figure 10.8 for the case where it is managed by an external party.

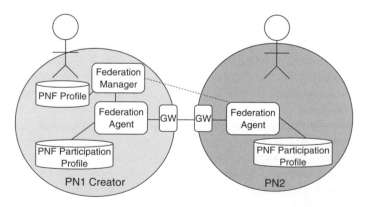

Figure 10.7   Main architectural components of a PNF when managed by one of the members.

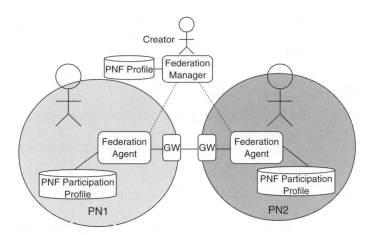

Figure 10.8 Main architectural components of a PNF when managed by an external party.

The entity that takes care of the management of the PNF is the *federation manager*. The federation manager may reside in one of the PNs or could be a service provided by a third party. In the PNF formed by Jane and her colleagues for sharing photos and printing services, Jane's PN will take the role of federation manager if she is the one initiating the PNF. In other PNFs, the management may be performed by an external party, such as a traffic coordination center or a health care center in the road safety and assisted living examples, respectively.

The federation manager takes care of controlling the access of potential members to the federation as well as issuing membership credentials to the participants. It also provides a service directory with an up-to-date list of the services available in the PNF to the members. Coming back to the example of Jane, the service directory will indicate the availability of a camera phone, a laptop, a computer with photo editing software, a digital photo camera, a photo printer, and a digital video camera as well as their addresses.

The federation manager contains a database where the *PNF profile* is stored. The PNF profile is bound to the entire federation and global information relevant to every member of the federation. The PNF profile is divided into a generic part, which is valid for the whole PNF, and a member-specific part. The generic part contains information about the PNF, such as its name (e.g. 'Federation for sharing conference material'), objectives (e.g. 'sharing, exchanging, and printing of conference material'), the rules for participation (e.g. who is allowed to join), the resources and services that are required (e.g. every PN of colleagues or friends in Jane's list of contacts), and a list of PNs that have joined (e.g. PN2, PN3 and PN4). The member-specific part specifies the services that are made available to the PNF and the mutual agreements of the provisioning of those services.

In addition, each participating PN has an individual *PNF participation profile* that defines the resources and services that the PN is prepared to make available to the federation (e.g. photo camera and photo printer for PN3). Each PN has a *federation agent* that manages its participation in the federation and coordinates and controls the access to the services and resources that are temporarily committed to the federation.

As mentioned in Chapter 7, PNs communicate with other PNs through their gateway nodes. This also holds true for PNFs. Hence, a PN communicates with the other PNs in a PNF through one or more gateway nodes.

10.5 Life Cycle of a Federation

The user or provider initiating the federation is called the *creator*. The creator has the responsibility of providing the federation management services.

A federation has a temporal nature with a life cycle, which is illustrated in Figure 10.9 and comprises the following five phases:

Initialization phase. Here, the PNF profile and the PNF participation profile are defined by the creator.
Discovery phase. This phase consists of the publication of the PNF profile by the federation manager and the discovery of potential federations to join by the potential members.
Participation phase. In this phase, the group of participants is built up.
Operation phase. In this phase, the members of the federation can share resources and services.
Dissolution phase. This phase is either initiated by the federation manager or is the consequence of specific conditions.

The above phases are discussed in the following sections in more detail.

10.5.1 Initialization

During the initialization phase, the creator, via the federation manager, defines the PNF profile. This profile is common to the federation and contains the policies, membership management rules, and additional information necessary for its formation. The profile may contain information on how and when a federation can be formed, the minimum number of participants, the lifetime of the federation, and criteria for its termination.

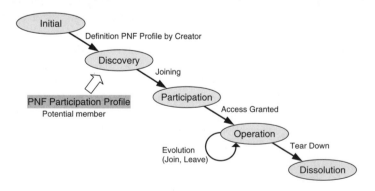

Figure 10.9   The life cycle of a PNF.

Every member determines individually which resources it wants to make available to the federation in their PNF participation profile. Besides a list of accessible services, the participation profile contains information about the member, such as PN identity and contact information.

In addition to the definition of the profiles, the federation manager generates a public–private key pair and a root certificate for the PNF in the initialization phase.

10.5.2   Discovery

During the discovery phase, the federation manager advertises the profile of the PNF so that it can be discovered by potential members. In the case of an infrastructure-supported federation, the PNF profile can be advertised by a federation broker. The federation broker can also store membership profiles of potential users and help to discover a candidate PNF that matches their interests. In the case of an ad-hoc PNF, the profile could be advertised by periodic broadcasts via one or more communication interfaces. After a successful discovery, a candidate member must contact the federation manager to join the PNF.

10.5.3   Participation

In this phase, the negotiations for joining the federation take place as well as the authentication and authorization processes and the establishment of the security associations with the participants for securing their further communication. The participation phase starts with the mutual authentication of the PN member and the PNF manager. In an infrastructure-based PNF, the identities can be certified by a registered trusted third party (TTP). In the ad-hoc case, a proximity-based authentication mechanism could be used. Security is discussed in more detail in Section 10.8.

The authentication method is specified in the PNF profile defined by the creator. After a successful authentication, granting access to the PNF will depend on the concrete policies and rules defined in the PNF profile and the PNF participation profiles. The services that are made available to the PNF and the conditions for their use may depend on the mutual trust between the new participant and the creator. The rules may include the reputation of the other parties.

A participant having been granted access to a PNF, the federation manager stores the participation profile of the new participant in its PNF database. The database contains a list of all the members participating in the federation, together with their identity, point of contact (e.g. IP address), participation profile and service list (MAGNET Beyond 2007).

10.5.4   Operation

Once a PN has become a member of the PNF, it is allowed to access services provided by the other members. During the operation phase, new members can join while other participants can leave. This makes the PNF dynamic, because its members and available services change over time.

In order to locate a specific service, a PN contacts the federation manager, which provides service discovery functionality. When a member requests a specific service from a PN, the federation agent checks in its own participation profile whether the requested service is allowed within the context of the current federation. If the service is allowed, then the federation agent sends back the location of the service.

10.5.5 Dissolution

This phase may be initiated by the federation manager, with all parties being given a chance to react before the PNF is dissolved, or it may take place abruptly as a result of specific conditions.

The federation manager may start the dissolution of the PNF for a number of reasons: the objectives of the federation may have been met, insufficient resources are available for satisfactory operation, or the impossibility of reaching the goal within the specified time. An abrupt dissolution may be caused by conditions such as all participants having left the federation, or connectivity with the federation manager having been lost.

In the dissolution phase, all security associations are cleared, the members are informed and (optionally) the history of interactions of the members is stored for future use.

10.6 Federation Access Control

The access control in PNFs takes place at two levels. *First-level access control* takes place when a new member wants to join the PNF. It is carried out by the federation manager. *Second-level access control* takes place when a member of the PNF wants to make use of a service provided by another participant. The access control at this level is performed by the federation agent of the PN where the service is located.

10.6.1 First-Level Access Control

First-level access control is the mechanism for accepting or rejecting a potential participant. It takes place after the discovery phase and initiates the participation phase. Figure 10.10 shows the basic steps that take place.

The participation phase starts with the mutual authentication of the federation manager and the potential member. Different ways of authentication are possible, but the authentication method for a specific PNF is defined in the PNF profile. After a successful authentication, the federation manager performs the authorization procedure. The authorization procedure determines whether the potential participant is allowed to become a member of the PNF or not. Access may be granted or denied depending on a variety of criteria expressed in the form of policies. If access is granted, the new participant will receive access credentials that should be verifiable by the other participants. The credentials may be in the form of membership credentials digitally signed by the federation manager.

10.6.2 Second-Level Access Control

Second-level access control takes place within the operation phase of the PNF and is performed at the PN level when another participant wants to make use of one of its services. The steps involved are shown in Figure 10.11.

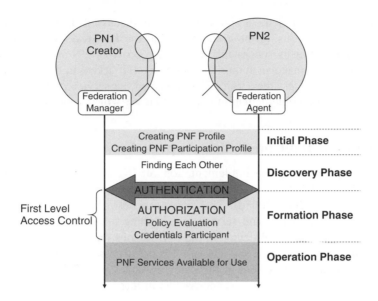

Figure 10.10   First-level access control.

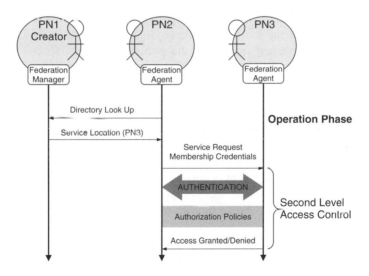

Figure 10.11   Second-level access control.

The participant requesting the service learns about the point of contact (e.g. IP address) of the PN that provides a service by using the service lookup provided by the federation manager. Once the point of contact is known, the service-requesting PN sends a service request to the service provider PN. At this point, the second-level access control procedure is initiated. The procedure takes place either at the gateway node or at each node, depending on the implementation approach, as discussed in the next section. The authentication takes place using the credentials that the PNs received when joining the PNF. If the authentication is successful, the service provider PN will apply the service access control polices to make an authorization decision. If access to the service is granted, the requester is able to access the service. The way that the service provisioning is carried out also depends on the implementation choices that we will discuss in the next section.

10.7   Federation Implementation Approaches

There are two approaches to the implementation of PNFs: one using network overlays between PNs (MAGNET Beyond 2008b) and one using service proxies at the gateway nodes of the PNs (PNP2008 2008d). The difference between the two approaches is in the way service access control and service provisioning are carried out. In the *network overlay* approach, each personal device in the PNF carries out the service access control. In the *service proxy* approach, the services of a PN are accessed by other PNs not directly at the service providing device, but at the gateway nodes of a PN by means of service proxies. We discuss each of these two approaches in turn.

10.7.1   Network Overlay

The network overlay approach is based on establishing a virtual network with all the PN devices and services that the participants make available to the PNF. The PNF uses a virtual address space separated from both the public IP address space and the address spaces of the participating PNs to confine the communication within the scope of the network overlay.

After the participation phase, all the gateway nodes of the participating PNs proactively create tunnels from one gateway node to all other gateway nodes. When a service is requested, the result of the service discovery is the location of the service (node IP address in the PNF address space). At the service provider PN, the component involved is prepared to accept the forthcoming service session establishment. On the requester side, the address of the service component is passed to the client. Using the address, the client sets up a service session with the service component. The request procedure includes the necessary second-level access control procedures. Figure 10.12 shows the basic idea of this approach.

In the network overlay approach, all the nodes of the services and resources that are committed to the PNF participate in the network overlay. Although the authentication is done at the gateway nodes of the PNs, service access control is carried out at the nodes hosting the services. Therefore, the overlay approach uses a distributed access control for the PNF resources.

The advantage of this approach is that once the overlay is established, service access is fast. The drawback is that additional functionality is required at all service hosting nodes. In addition, the number of connections to be maintained grows exponentially when new services are added to the PNF, raising the complexity of service management.

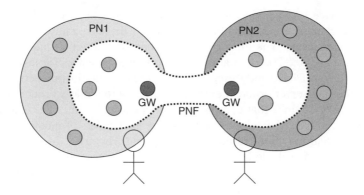

Figure 10.12   Network overlay approach to federations.

10.7.2   Service Proxy

The service proxy approach to the PNF architecture is based on the use of service proxy components at gateway nodes of the PNs. We introduced this concept for PN foreign communication in Chapter 7. For PNFs, too, the service proxies keep the services and resources and other internal structures of the PN hidden from the other participants in the PNF. In this way, it becomes possible to separate the security mechanisms and other mechanisms between the PNF and the PNs.

In this approach, the access control of the PN services is carried out not in every node in a PN, but at the border of a PN (i.e. the gateway nodes), so other personal nodes inside the PN need not have access control capabilities. Having the access control at the borders allows each PN to have a separate security domain in a PNF and to keep its autonomy.

Figure 10.13 shows the basic idea of service provisioning in a PNF using service proxies. PN1 is the service-requesting PN, whereas PN2 is the service-providing PN. The service proxy in PN2 acts as a server for the service proxy in PN1. To the client in PN1, the service proxy in PN1 appears as the server. To the actual server in PN2, the service proxy in PN2 appears as a client.

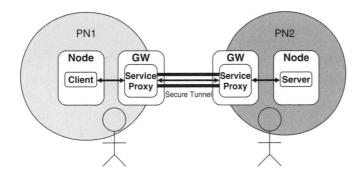

Figure 10.13   Service provisioning using the service proxy approach.

For secure service provisioning across public infrastructures or other untrusted networks, a secure channel can be established between the two gateway nodes where the two service proxies reside. The tunnels can be established at either the network layer by using IPsec (Kent and Seo 2005), or at the transport layer by using, for instance, Transport Layer Security (TLS) (Dierks and Rescorla 2008). The communication between the service proxies and either the server or the client is protected by the internal security mechanisms of the participant PNs, since that communication is entirely within the respective PNs.

In its simplest form, a service proxy just changes the headers of incoming and outgoing packets without doing anything further to the packets. However, in many cases, it may be interesting to provide a modified version of the services, such as providing only a subset of the whole functionality. In such a case, the service proxy is allowed to modify not only the headers, but also the payload of the packets. The components in the service proxy that change the payload are called *service handlers* (PNP2008 2008d). The way a specific service handler deals with packets of a connection depends on the type of service and the authorization policies. Observe that a service handler may not be required for all services.

As also mentioned in Chapter 7, the fact that service access control and provisioning are carried out at the borders has the advantage that each PN has centralized control over its resources and services, relieving the nodes inside the PN. Consequently, even a simple device in a PN can export its services with the support of a service proxy. In addition, the personal nodes in the PN are not directly accessible by the other PNF participants, allowing each PN to keep its autonomy by having a separate security domain from the security domain of the PNF. The main drawback of this approach is the extra processing load at the gateway nodes, which can act as a performance bottleneck if the number of services used increases. However, a hybrid architecture that combines the advantages of the network overlay and service proxy approaches has been proposed in (Ibrohimovna and Heemstra de Groot 2009).

10.8   Security

Obviously, security is essential in the PNF concept. From the point of view of security, we need to discuss how trust and security associations are set up between the creator and the PNF members as well as among the PNF members (MAGNET Beyond 2006b).

10.8.1   Trust between the Creator and a New Member

Depending on whether or not the PNF makes use of services in the infrastructure for its establishment, we can distinguish between infrastructure-based and ad hoc PNFs. This distinction is essential when establishment of trust is based on identities. Infrastructure-based federations can make use of solutions that involve the services of a TTP to provide certified credentials, whereas ad-hoc federations require the direct involvement of the users. We discuss these two scenarios in more detail below.

Infrastructure-based PNFs. This scenario assumes that all the PNs have access to a
TTP that provides means for authenticating the potential participants. This implies a
prior enrollment in the TTP by all the PNF members.

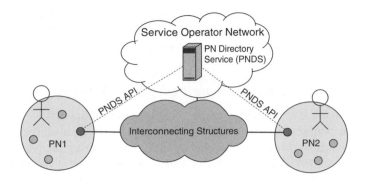

Figure 10.14 PNDS as a trusted third party (Alutoin et al. 2007).

An example of a TTP for facilitating authentication between PNs and building PNFs is the Personal Network Directory Service (PNDS) (Alutoin et al. 2007; MAGNET Beyond 2008a). The PNDS is based on a business model in which the users register with a service provider and establish their federations through it. The PNDS stores public keys and provides PN certificates for those public keys. It also acts as a directory service for federations. It is based on the existence of a common TTP for all PNF participants, but can be easily extended to the general case of hierarchical certificate authorities. A high-level view of the PNDS is shown in Figure 10.14.

There are basically two ways to use the directory services to establish an infrastructure-based PNF. In the publish mode, the creator uses the PNDS to advertise the PNF profile, while potential members look in the directory for suitable PNFs to join. In the invitation-based mode, potential members announce to the PNDS their willingness to participate in specific PNFs and the creator searches in the directory for suitable candidates to invite. Both solutions require the same level of trust. After an invitation by the creator or a request to join by a candidate member, the creator and the candidate authenticate each other based on their PNDS certificates.

Ad hoc PNFs. Ad hoc federations do not rely on services provided by TTPs for their formation. The trust relation has to be built up with the involvement of the PN users. If the users could come into close proximity of each other, they could exchange passwords or let their devices make use of a location limited channel (LLC) (see Section 9.1.2) for an authenticated exchange of their public keys. This bootstrapped information can be used later for authenticating each other.

10.8.2 Security Association between the Creator and a New Member

Independently of whether the PNF is infrastructure-based or ad hoc, a security association between the PNF creator and the new member is created if the trust establishment phase is successfully completed. For this, they can use any well-established public key based key exchange protocol such as Transport Layer Security (TLS) (Dierks and Rescorla 2008). The establishment of a secure channel allows secure negotiations on the conditions to join the PNF. After agreement, the new member transfers its participation profile with the resources and services that are made available to the federation. Having granted access to the new member, the creator forwards the private part of the PNF profile and a group key to the new participant using the secure channel.

10.8.3 Security Association among Federation Members

Different mechanisms can be used to create a security association among the members of a federation. The simplest way is the use of a shared group key for all the members. Such a key could be distributed as part of the private PNF profile during the participation phase. PNF participants authenticate each other by showing knowledge of the shared group key and may use this key for secure communication within the federation. This allows a PN to be authenticated as a legitimate member of the federation, but it does not provide the means to identify individual members. Moreover, if a member leaves the group or is expelled, the creator needs to provide a new group key.

A more convenient solution for larger PNFs is to let the creator be the certification authority for the PNF. After successfully joining the PNF, each participant obtains a certificate on their authenticated public key. The certificate has validity only within the scope of that PNF.

The certificates are used by the participants to prove their membership of the PNF and to derive security association among the members. Certificates allow the identification of individual participants. Revocation of members is possible by implementing certificate revocation lists (CRLs) containing the list of revoked certificates that have not expired.

10.9 Summary

Not only bilateral communication between PNs is of interest – group communication among several persons and their PNs is also of importance. In PNFs, the PNs will interact and collaborate for the purpose of achieving a common goal. A PNF may host cooperative services by using shared resources from the participating PNs. The PNF is an example of pervasive and ubiquitous computing technology that builds on top of the PN concept. PNFs enable a selected subset of personal services to be exported to other parties, thereby enhancing the individual capabilities.

PNFs may have applications in many different areas, such as education, health care, entertainment, public safety, and business. Different subsets of personal devices and services of a single PN may become available to a large variety of PNFs, each with its own access policies and membership rules.

In this chapter, we have presented the main principles of PNFs as an extension of the basic PN technology. We discussed the basic architectural components, its life cycle, and the access control and security framework. Although PNFs may benefit from functionality provided by a fixed infrastructure, as certification authorities, directory and broker services, the paradigm also supports ad hoc scenarios.

11

Personal Network Prototypes

At the time of writing this book, there are at least three PN prototypes. In this chapter, we briefly introduce them and highlight the differences. We explain the TU Delft prototype in detail. In doing so, we demonstrate how the concepts and mechanisms introduced throughout this book can be implemented. The other two prototypes are the prototypes developed by the PNP2008 project (http://pnp2008.freeband.nl/) and the MAGNET project (http://magnet.aau.dk/).

The PNP2008 project was funded by the Dutch Ministry of Economic Affairs under the Freeband Communication Technology Program and ran for just over four years. Within PNP2008, several different PN prototypes with different purposes were developed. The early PNP2008 prototypes were designed to demonstrate the usefulness of PN technology, but using only state-of-the-art technology. As the work progressed, attempts were made to make more complete PN prototypes based on the PN technology developed.

MAGNET and its continuation, MAGNET Beyond, were European research projects funded by the European Commission. They had around 32 partners from 15 countries, including highly influential industrial partners, universities, and research centers. The projects were an attempt to research and define PN concepts using experts from many different fields. The aim included investigating business models for PNs, PN user needs and requirements, PN networking, PN security, new wireless communication technologies for PNs, and an integrated PN prototype. The MAGNET prototype was far-reaching and also included hardware implementations of the new wireless techniques.

This chapter is intended mainly for those interested in the technical details of how PNs can be implemented in software. To be able to follow this chapter, the reader needs to be familiar with the content of Chapters 4–10. Furthermore, a good knowledge of Linux and its network implementation is recommended.

This chapter is organized as follows. In Section 11.1, we start by introducing the TU Delft prototype in detail. Sections 11.2 and 11.3 continue by introducing the prototypes developed by the PNP20008 and MAGNET projects, respectively. Finally, we summarize the chapter in Section 11.4.

Personal Networks: Wireless Networking for Personal Devices   Martin Jacobsson, Ignas Niemegeers and Sonia Heemstra de Groot
© 2010 John Wiley & Sons, Ltd

11.1 The TU Delft Prototype

The purpose of the TU Delft prototype was to demonstrate the networking concepts of PNs and to test out certain mechanisms. It is based on the Linux operating system and builds on open source software. This section introduces the prototype implementation and discusses some findings. It can serve as an example of how PNs can be implemented in real products.

The TU Delft prototype implements many of the networking features of a PN, such as:

(i) neighbor discovery using network level hello messages for cluster formation;
(ii) link quality assessment (LQA) based on received signal strength and data packet retransmissions;
(iii) intra-cluster routing based on OLSR and using the LQA information;
(iv) inter-cluster tunneling and routing based on a central PN agent, but with direct routing, mobility support, and basic encryption.

11.1.1 Hardware Platform

The prototype was implemented on six standard laptops. This allowed us to avoid cumbersome development processes, common when using smaller devices, such as embedded software, cross-compilers, special debuggers, etc. The six laptops were equipped with Ethernet and WLAN based on IEEE 802.11 (IEEE 1999). Two laptops had an Intel Core 2 Duo (1.66 GHz) processor, two had an Intel Celeron M (1.6 GHz) processor, and two had an Intel Mobile Pentium 4 (3 GHz) processor. All laptops had 512 MB RAM. For the wireless functionality, we mainly relied on the 3Com OfficeConnect Wireless 108 Mbps 11 g XJACK PC Card (http://www.3com.com/products/en_US/detail.jsp?tab=features&pathtype=purchase&sku=3CRXJK10075), but also on the built-in Intel PRO/Wireless 2200BG WLAN card available in some of the laptops. The support under Linux for these wireless interfaces was good.

11.1.2 Software Platform

This prototype was built on Linux. Linux (http://www.kernel.org/) is a free open-source operating system and runs on many different hardware platforms, including PCs, laptops and PDAs. It provides everything that can be expected from a state-of-the-art operating system. Since the source code is available, it is possible to adapt the system as required to build a cluster node prototype.

Linux provides a virtual network interface in software called Ethertap (http://vtun. sourceforge.net/tun/). To the rest of the system, Ethertap looks like a normal Ethernet interface except that it does not physically exist. We used the Ethertap interface to implement a virtual interface for intra-cluster communication. In this way, no modifications were needed to the kernel or the end-user applications. Instead, it allowed us to implement everything in a normal user space process. We could create a user space process that connects to the virtual Ethertap interface, which is done through a special device file. When a packet is forwarded to the interface, it is handed to that process, which can then process it. In a similar way, the process can create packets and send them to the virtual interface, which will forward them to the kernel in the same way as when a packet is received on a real interface. Ethertap was originally written to support user space implementations of tunneling protocols.

The Madwifi (http://www.madwifi.org/) driver is an open source driver for wireless cards based on the Atheros chip for Linux. This driver was chosen because its entire code is available, and because it implements many features usually implemented in firmware and running on the wireless card itself. This meant that we had access to the rate adaptation and link layer retransmission functionalities. In particular, this driver allowed us to implement LQA mechanisms that are based on received signal strength and on feedback of the retransmissions. For our experiments, we used Madwifi version 0.9.3.2 in combination with the 3Com OfficeConnect Wireless 108 Mb 11 g PC Card.

For the routing, we used OLSRd (http://www.olsr.org/), which is a Linux implementation of the Optimized Link State Routing Protocol (OLSR). Since Ethertap emulates a real network interface, OLSRd could operate on top of the virtual interface with no modifications. However, we also wanted to use the LQA information for routing and hence made some modifications to OLSRd to use that information.

For the implementation of the inter-cluster signaling protocol, we used Python (http://www.python.org/), an object-oriented script language that has a very rich set of built-in libraries. Python is available on all major platforms, including Linux. It is an easy-to-learn language that possesses most of the functionalities that can be expected from a modern programming language, such as a clean syntax, exception-based error handling, and high level dynamic data types. However, perhaps Python's best characteristic is that it allows one to quickly implement complex software, and that was the reason why we chose Python. Since it is a script language, it may not give us the fastest implementation, but should still be able to perform well.

For the encryption parts, we used libmcrypt (http://mcrypt.hellug.gr/lib/index.html), mainly because it offers an extension module for Python. Since libmcrypt itself is not implemented in Python, but in C, these parts of the code are still very fast.

11.1.3   Intra-Cluster Implementation

With this prototype, no changes are required to the applications; they send and receive traffic in the normal way. The intra-cluster network layer is based on IPv6 (Deering and Hinden 1998) and currently supported network interface types are fixed Ethernet and WLAN according to the IEEE 802.11 family. Node discovery is implemented at the network layer and is based on UDP packets and link local multicasting typical of IPv6. The various packet formats are shown in Figure 11.1.

To fully prototype a working cluster, addressing and routing are also needed. The addressing was done manually, but could be done automatically as part of the personalization functionality. A special prefix (temporarily 3000::/16) was assigned for all intra-PN addresses.

The link layer security features of the supported network types are not used. Instead, we show the feasibility of implementing the intra-cluster communication mechanisms entirely at the network level for both fixed Ethernet and WLAN. Data packets are encapsulated in UDP and sent with IPv6 packets using link local addresses as shown in Figure 11.1. Obviously, it is unnecessary to encapsulate data packets like this; it would be better to

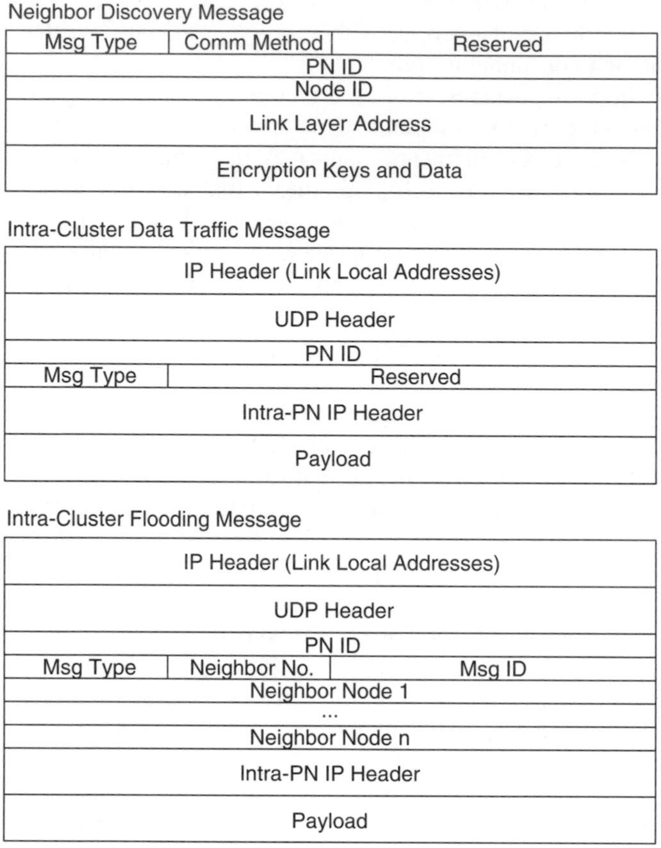

Neighbor Discovery Message

Msg Type	Comm Method	Reserved
PN ID		
Node ID		
Link Layer Address		
Encryption Keys and Data		

Intra-Cluster Data Traffic Message

IP Header (Link Local Addresses)	
UDP Header	
PN ID	
Msg Type	Reserved
Intra-PN IP Header	
Payload	

Intra-Cluster Flooding Message

IP Header (Link Local Addresses)		
UDP Header		
PN ID		
Msg Type	Neighbor No.	Msg ID
Neighbor Node 1		
...		
Neighbor Node n		
Intra-PN IP Header		
Payload		

Figure 11.1   The intra-cluster prototype packet formats.

introduce a new network protocol that only has a short header and whose purpose is to encrypt the entire intra-cluster communication packet. However, we made this choice since it makes it easier to implement.

Figure 11.2 shows a schematic view of the prototype implementation of a node without gateway node functionality. Thick arrows denote data traffic and thin arrows denote control, routing, and device discovery traffic. The virtual interface implemented by Ethertap is called ppan1. It is the interface always used for intra-cluster communication. The program connected to ppan1 is called ppand. It is also ppand that implements the personal node discovery and authentication process. It maintains the personal node neighbor table (PNNT) and makes sure packets on the virtual intra-cluster interface (ppan1) are encrypted before being sent to a neighbor. Security functions have not yet been implemented in the prototype. Proper personal node authentication and hop-by-hop encryption require some extra message exchanges and headers for key negotiations and more, but otherwise the implementation will be similar.

It is worth pointing out that data traffic actually passes through the kernel routing module twice. First, all intra-cluster traffic is passed to the virtual intra-cluster interface,

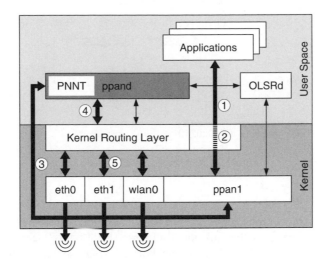

Figure 11.2   The cluster prototype implementation architecture.

and then ppand adds a second IPv6 header with a link local address of a direct neighbor personal node. Before the packet is sent to the next hop, it passes the routing table once more. However, our performance tests indicate almost no degradation because of this.

11.1.4   Sending and Receiving Intra-Cluster Traffic

Let us go step by step through the process of sending a packet. Consider an application on node 3000::1 in Figure 11.3 sending a packet to node 3000::3. Following the steps in Figure 11.2, this is what takes place:

1. The application sends the packet to an intra-cluster address (3000::3). The packet will be delivered to the kernel through a `send()` call and the routing layer in the IP stack will be called.
2. The routing layer will find the next hop IP address, which is also an intra-cluster IP address. The next hop IP address is bound to the virtual intra-cluster interface (`ppan1`) and will therefore be delivered to ppand, but before the routing layer can deliver it, it needs the link local address of 3000::2 on `ppan1`.
3. The kernel will then try to find the link layer address (Ethernet address when using Ethertap) for 3000::2 on `ppan1`. A neighbor discovery packet (ICMPv6) is first sent to `ppan1` and delivered to ppand. Here, ppand answers this request with a unique address. Any address will do as long as it is unique on `ppan1` and always the same for 3000::2. The kernel will then deliver the packet to ppand using that link layer address as link layer destination and 3000::3 as the IPv6 destination. When ppand receives this packet, it knows both the next hop (from the link layer address) and the final destination (from the IPv6 address). This process taking place between the routing layer and ppand can be seen as a pure internal mechanism and can be optimized in a real implementation.

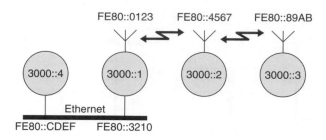

Figure 11.3   An example scenario.

4. The ppand has a PNNT, which is shown in Table 11.1. The PNNT should contain the next hop node. To encrypt the packet, ppand will use the pairwise key of the next hop (3000::2). It also extracts the link local address (FE80::4567) and interface (wlan0) of the next hop node and sends it to the kernel using sendto() encapsulated as in Figure 11.1.
5. The kernel believes the packet is for the next hop, so it forwards it to the correct real interface for transmission as usual.

Table 11.1 shows the PNNT of node 3000::1 and Table 11.2 shows the routing table of node 3000::1 in this scenario. The PNNT contains only personal neighbor nodes and maps PN-internal addresses to external addresses. In this prototype, we use the IPv6 link

Table 11.1   Personal neighbor node table (PNNT) of node 3000::1.

Personal Node Address	Current Link Local Address	Interface	Pairwise Key
3000::2	FE80::4567	wlan0	90ab74bf582b3e28c
3000::4	FE80::CDEF	eth0	a8493eb57e7f43cc8

Node 3000::1 has two neighboring personal nodes. For each node, the interface used, the link layer address of the node, and encryption keys are kept in the table.

Table 11.2   The routing table of node 3000::1.

Destination	Next Hop	Interface
3000::3	3000::2	tap0
3000::2		tap0
3000::4		tap0
FE80::/64		wlan0
FE80::/64		eth00

Node 3000::1 has routes to three nodes in its own PN. This means that it belongs to a cluster of four personal nodes. Two of the nodes are neighbors, while one node is two hops away, as shown in Figure 11.3. The last two entries are standard route entries for IPv6 devices.

local address of the node. In a real implementation, the medium access control address is a better option. We also need to keep the pairwise key for encrypting/decrypting the data traffic to and from the neighbor node.

Ppand binds to the UDP port used for intra-cluster traffic and receives all intra-cluster packets for processing. When an intra-cluster packet arrives from a neighboring personal node, it looks at the source address and identifies the neighboring personal node and its pairwise key in the PNNT. Ppand then tries to decrypt the content and if successful, sends the decrypted packet to `ppan1` for further processing by the kernel. The kernel routing table decides whether the packet is to be sent to one of its applications or whether to forward it to another node. In the latter case, the packet is forwarded back to `ppan1` and ppand for encryption and transmission to the next hop.

11.1.5   Interface Output Queue

It is necessary to give control packets, such as hello packets and routing packets, higher priority when a queue arises on the intra-cluster interfaces. This is done on the output queue of each of the interfaces using the Linux traffic control feature (Almesberger 1999). The traffic control decides which packet to send first, whether to delay a packet transmission, which packets should be dropped, etc. The default output queue on Linux is a first in first out (FIFO) queue with tail drop, but can be changed by a configuration tool called `tc`, available in the iproute2 package. With `tc`, advanced queuing structures can be set up for each outgoing interface.

To make sure hello packets and routing packets are never delayed or dropped due to full buffers, we introduced a priority queue on each interface used for intra-cluster communication. The priority queue had only two priorities: high, which is always processed first, and low. Filters were installed that made sure hello packets and routing packets got high priority. Since both routing and data packets are encrypted, we simply gave all broadcast packets high priority since all routing traffic uses broadcast.

11.1.6   Intra-Cluster Flooding

We also wanted to know how the flooding protocols described in Section 5.4.4 can be applied to intra-cluster communication. Therefore, we extended our PN prototype with four different flooding protocols. The flooding algorithms were implemented in ppand itself. To allow any application to use optimized flooding without modifications, we let packets with an intra-cluster multicast destination address with a prefix of FF13::/16 be treated as flooding packets by our flooding code. Packets with other multicast destination addresses use the normal link local multicast mechanism only.

To implement cluster-wide flooding, we needed to make some small additions to the intra-cluster data traffic message. The new fields include a message identifier, a neighbor node list, and the number of nodes in that list as shown in the last part of Figure 11.1.

The system distinguishes packets from each other by using the Msg ID field. A packet is uniquely identified by the PN ID, node ID, and Msg ID. To make sure a receiving node does not process a packet twice, it keeps a list of recently received packets and compares new packets it receives to packets in the list. If a match is found in the list, a received packet is handled as a duplicate, otherwise as a new flooding packet.

The actual logic for the flooding algorithm was implemented in ppand. Hello message information was taken from ppand's PNNT table, which was extended to also contain link quality information specifically for flooding. Four flooding protocols were implemented in ppand: Blind Flooding, Counter-Based Broadcasting (CBB) (Tseng et al. 2002), Prioritized Flooding with Self-Pruning (PFS), and Counter-Based PFS (Jacobbson et al. 2005a).

11.1.7   Intra-Cluster Routing

OLSRd was set up to handle the intra-cluster routing by being configured to send and receive its routing packets only over the virtual intra-cluster interface. This has the effect that all OLSR messages are encrypted and protected by ppand and its security system. Further, only neighboring personal nodes can receive and decrypt the OLSR messages. As a consequence, OLSRd will only see personal nodes in its tables. Hence, there is no need for special modifications to the routing daemon, at least not for intra-cluster routing.

OLSRd directly updates the kernel routing table. Each destination within the cluster was pointing to the virtual intra-cluster interface, but with different next hop addresses. The discovery or disappearance of neighbor nodes will add or remove visible nodes from the virtual intra-cluster interface. As soon as this happens, the routing daemon can send and receive hello messages to and from these neighbors and detect topology changes. Also broadcast or multicast packets sent to the virtual intra-cluster interface are only received by neighboring personal nodes. However, since ppand implements a personal node discovery protocol, it is unnecessary for OLSRd to have its own. It is better if ppand informs the routing daemon directly when a node is discovered or disappears. Therefore, a special inter-process communication (IPC) mechanism was designed to connect ppand and the routing daemon. With this IPC mechanism, OLSRd got access to ppand's PNNT and the associated link quality information.

To improve the link quality information gathered, as explained in Section 5.4.2, we extended ppand to enable proper LQA and high quality unicast routing within a cluster in our prototype. In addition to standard hello packets, the LQA code periodically reads the received signal strength and the feedback from the retransmission mechanism. All this raw data is fed to a function within ppand that calculates the LQA, which is then sent via an IPC mechanism to the routing daemon.

The OLSRd routing daemon already implements OLSRv1 (Clausen and Jacquet 2003) including a link quality extension based on expected transmission count (ETX) (De Couto et al. 2003a) using hello packets. To use the LQA values from ppand, we enabled this link quality extension, but replaced it with our own functionality that retrieves the LQA values from the ppand. For simplicity, OLSRd still sent and handled its own hello messages (link quality-enabled hello messages) as usual. However, the reception (or non-reception) of the hello messages was only used to determine the neighbor set, but not the link quality. The link quality information came from ppand via the IPC mechanism, as explained earlier.

11.1.8   PN Organization

We implemented both the PN agent and the gateway node functionality for inter-cluster communication. In contrast to the intra-cluster communication, we did implement encryption and data integrity protection for the inter-cluster communication and signaling. While the inter-PN traffic is based on IPv6 as usual, only network access points and interconnecting structures using IPv4 are currently supported. All inter-cluster communication and signaling code was implemented in Python.

The central part of both the PN agent and the gateway node implementations is the PN database that holds information about known and relevant gateway nodes and their active tunnel endpoints (TEPs), as explained in Chapter 6. Figure 11.4 shows the design (in the Unified Modeling Language, UML) of this database, which we call PNDB. What is shown

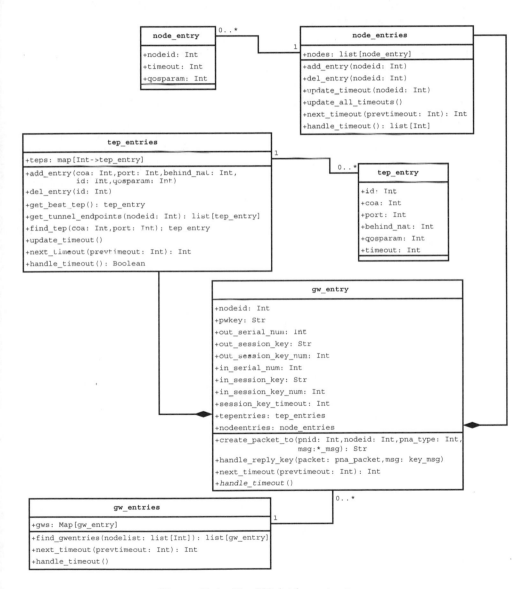

Figure 11.4   The PN database structure.

are the common parts shared between the PN agent and gateway node, including all the kept states. We can clearly see the various entities, their attributes, and how they relate to each other. The implementation of PNDB is exactly as shown in Figure 11.4, which means a lot of sequential searches through the database. Obviously, a scalable solution would need to use more efficient data structures. Nevertheless, this implementation still performed well for our test PNs.

The PNDB consists of a list of all known gateway nodes and their pairwise keys. For each gateway node, the PNDB may hold its active TEPs and the personal nodes belonging to its cluster. As explained earlier, this data is complete in the PN agent's PNDB. For the PNDBs in the gateway nodes, this information may be partial, but if present for one particular gateway node entry, it is complete and updated for that entry.

For the inter-cluster communication and signaling protocol, we defined one common packet format that can carry three different message types. The key message carries security key information, the TEP node list message carries TEP and cluster member node information, while the last message type carries encrypted user data. Figure 11.5 shows all the packet and message formats in detail.

The encryption is modeled after IPsec ESP but somewhat simplified. The key messages are encrypted and signed with the pairwise keys. They are used to agree and communicate the session keys between two gateway nodes or between a gateway node and the PN agent. All other messages are encrypted using those session keys. While this security implementation captures the most essential aspects of security, it may not be the best option. Other, more advanced schemes, such as the ones outlined in Chapter 9 above, MAGNET (2005c), MAGNET Beyond (2006b), or Jehangir and Heemstra de Groot (2007), should be considered for implementations in future PN products as they may offer better security and performance.

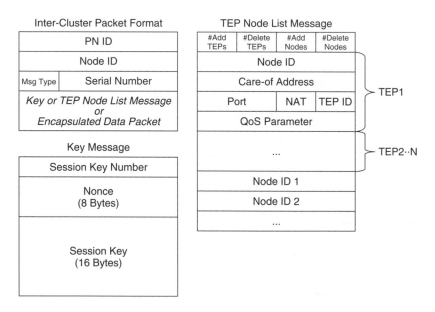

Figure 11.5 The inter-cluster signaling packets.

The TEP node list message has four lists that contain new TEPs, removed TEPs, nodes newly arrived at the cluster, and nodes that have left the cluster. The message format is flexible enough to be used for all message types regarding TEPs and cluster nodes. It can be used for TEP updates, cluster node updates as well as queries and update subscriptions for such information. However, not all lists are used in all message types.

More message formats can be defined. Examples include new message types regarding NAT traversals and data packet forwarding through a third party. However, we have not yet specified such message formats.

The PN agent implementation is actually very simple. It only consists of the PNDB with minor extensions plus functionality for listening on a UDP port and handling received packets. The implementation is completely event-driven and handles two types of events: timeout events in the PNDB and the reception of a UDP packet. Timeout events make sure that the PNDB stays up to date and that stale information is cleaned up. A received UDP packet is first parsed, its signature verified, and then decrypted. The message inside the packet is then completely parsed and handled. The PNDB is updated with the new information carried in the message, if any. Then the PN agent uses the information in its PNDB to formulate a reply or to inform other gateway nodes about the updates. Those messages are finally sent back encrypted using the session keys.

The gateway node implementation is much more complicated. The main difference is that in addition to keeping the PNDB and handling signaling packets, it needs to do data packet handling and information gathering regarding available local TEPs and the cluster node membership. Figure 11.6 shows how the implementation was done.

Except for some setup configuration scripts, all inter-cluster communication function-alities were implemented in one program named gwd. This is similar to ppand in the sense that it also creates a virtual interface (called pn1 in Figure 11.6) and opens a UDP port for communication with other gateway nodes and the PN agent. In addition to this, gwd also maintains the PNDB. It is responsible for all inter-cluster signaling as well as data packet encryption and tunneling. Further, it needs to interact with the operating system to retrieve information about available network connections and their status. This information flow, together with the inter-cluster signaling traffic, is indicated by

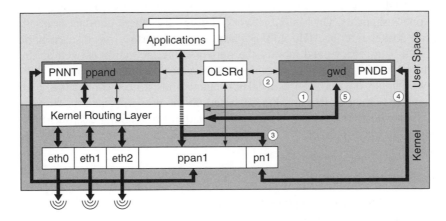

Figure 11.6 The gateway node implementation.

arrow 1 in Figure 11.6. Arrow 2 indicates the information exchange between gwd and the OLSRd routing daemon. The OLSRd routing daemon provides gwd with the cluster member node list, which it gathers from the intra-cluster routing messages. In return, gwd provides OLSRd with quality parameters regarding its current connections to the interconnecting structures. OLSRd propagates that information through the whole cluster so that every node can use the best-connected gateway node.

On a gateway node that has the best connection to the interconnecting structures, a data packet from an application on the node itself or from other nodes in the cluster (also other gateway nodes) will be forwarded to the pn1 virtual interface as indicated by arrow 3. The kernel will forward the packet to gwd (arrow 4) in the same way as it forwards packets arriving on ppan1 to ppand. If gwd has a valid tunnel that can be used for forwarding the packet toward its final destination, it will encrypt, encapsulate, and forward the data packet over the local interface associated with that tunnel according to arrow 5. The data packet is encapsulated in a UDP packet with the two TEPs of the tunnel as source and destination addresses and ports. If gwd does not have a tunnel that can be used, it will consult its PNDB for information on how it can establish such a tunnel. Since the PNDB may be incomplete, it may need to query the PN agent. In that case, the data packet will be enqueued by gwd until further information is retrieved and a tunnel can be established.

The gwd application does not discover connections to the interconnecting structures on its own, but expects the operating system to do so. This decouples the PN functionalities from the access network discovery technologies. It minimizes the impacts on existing legacy applications and also makes it possible for the PN communication to leverage the current and future developments in access networks.

Currently, most operating systems use DHCP (Droms 1997), IPv6 router solicitations and advertisements (Narten et al. 2007), DNAv6 (Krishnan and Daley 2009), or similar techniques for access network discovery. The most commonly used technique, DHCP, is typically handled by a user space program that reconfigures the routing table and other network-related settings in the operating system. Unfortunately, most DHCP clients will replace routes instead of keeping all alternatives. Because of this, we needed to slightly modify the DHCP client.

Further, gwd needs to be informed when a new access network is discovered or an existing one disappears. On Linux, network settings like these can be accessed through the netlink socket interface (He 2005). Through that interface, user space applications, such as gwd, can retrieve and modify the network interface table as well as the current routing table. Furthermore, it is possible to subscribe to update events for those tables, which gwd should do. As soon as a DHCP client finds a new connection, gwd is then updated and may then update its TEP list based on the new information. However, all of this, together with the detection of the quality of the discovered TEPs, has not yet been implemented in our prototype.

11.1.9 Lessons Learned

The CPU load of our test bed laptops remained at a moderate 10% during all experiments. Most of the computation concerned the extended logging facilities and the generation and measurement of test traffic. Even though we used Ethertap, which meant that data packets are passed between user and kernel space several times and one extra time through the kernel routing table, almost no performance degradation could be observed for intra-cluster communication when compared to communication without the PN prototype with Ethertap and all.

However, we did notice some performance degradation when using inter-cluster communication. The round trip time increased by 1.5 ms when measuring between two gateway nodes. Also the CPU load increased, which caused the maximum throughput to significantly degrade. A large part of the performance degradation can be attributed to Python, which had to handle not only signaling traffic but also every single data packet. However, since the per-data packet handling of gwd is not heavier than that of ppand in theory, the two should perform about the same if both were implemented in C. Hence, we strongly believe that an optimized gwd implementation in C will achieve performance similar to ppand. However, with encryption, the throughput dropped even further, while the delay increased by an additional 20 ms. Since the encryption implementation is done in C, this performance degradation cannot be attributed to the non-optimized implementation of the prototype. Instead, it demonstrates the importance of using hardware encryption.

11.2   The PNP2008 Prototypes

PNP2008 included prototyping right from the start. The early prototypes only used state-of-the-art technology, but were designed to experiment with and demonstrate the usefulness of PN technology. These prototypes were more focused on PN applications than on the PN technology itself. As the work progressed, the PN technology matured and, in the final PNP2008 prototype, attempts were made to make more complete PN prototypes. In this section, we introduce the earlier prototypes as well as the final prototypes.

11.2.1   Early PNP2008 Prototypes

The early PNP2008 prototypes, which are further covered in den Hartog et al. (2007), were designed to demonstrate and test potential PN applications. To enable the various PN applications, some support systems were developed such as basic content management and service discovery.

Some of the prototypes were used in user trials. The user feedback was used to expand the PN concepts and understand what is really needed. PNP2008 produced four application-centric PN prototypes:

Medicam. The first prototype was Medicam, which was a demonstration of PNs in a professional setting: the medical profession. It showed how PNs and PN federations can be used to easily tie devices together in areas where user mistakes are unacceptable. In the demo, a medical team was working in an operating theater equipped with high resolution video cameras, while another remote team could follow the operation using their computers. A kind of PNF based on invitations of the participating persons was formed using state-of-the-art technologies. The high resolution cameras were introduced into the PNFs as shared services. Also the control of cameras could be shared, allowing the remote team to control the pan, tilt and zoom functions.

To build the prototype, a whole range of technologies were used and manually integrated. The communication technologies used included GPRS, Bluetooth, and IEEE 802.11a.

Always at Home. The second PNP2008 prototype was developed to demonstrate the added value of PNs in the home environment. A user was equipped with a PN that included devices such as a PDA, an MP3 player, a wireless headset, a personal video recorder (PVR), a doorbell with camera, and a residential gateway. Furthermore, RFID technology was used to detect where in the house the user was.

Several features typically useful in a home environment were implemented. For instance, when the doorbell rang, context information, such as whether the user is at home or not and whether the PVR or the MP3 player are active, was used to determine how to alert the user. If the user was not at home, the video of the front door was streamed to the PDA, or, if the PVR was active, to the TV after pausing the PVR session. Using the TV or the PDA, the user could choose to open the door remotely. Other features included accessing music files on the residential gateway or the PDA from the MP3 player and pausing the MP3 player when there was an incoming call.

Again, state-of-the-art technologies were used and manually integrated. Due to a clear lack of common protocols and mechanisms, it was necessary for the developers to rely on many incompatible communication standards. The same problem was also encountered when enabling the use of a service on one device from the other devices.

MarcoPolog. The third PNP2008 prototype focused on user-generated content and location awareness for a traveling PN user. The PN consisted of a content server at the user's home and a couple of mobile devices that the user brought on the trip. The mobile devices, which included a PDA, a mobile phone, a GPS receiver, and an RFID reader, formed a cluster using Bluetooth. Using UMTS, the mobile phone could connect the cluster with the content server at home.

When the user took pictures, captured videos, or collected other electronic information, the content generated was temporarily stored on the mobile phone or the PDA. Then an intelligent file synchronizer uploaded file metadata as well as the content to the content server or other devices in the PN, based on the current connectivity. At the server, generated content was matched with the GPS coordinates and a map service to create a travel log showing where and when the user traveled and what content was collected at each location. The log could later be viewed by the PN user or shared with others.

This prototype demonstrated an excellent PN application that combined functionalities from several personal devices. Yet again, a lot of effort was required from the developers to set up and configure the PN with the travel log application.

PNPay. In the fourth PNP2008 prototype, the aim was to combine personal devices with external devices and services to enable intelligent travel planning and billing for a combination of public transport and personal transport. Using bus and train pricing and timetable information in combination with road pricing and road conditions (traffic congestions, road works, etc.), travel plans based on user preferences could be drawn up. In combination with devices deployed in public transport vehicles, the system was extended to also perform public transport payment, thereby avoiding the need to buy tickets, etc.

Some of these prototypes were used in real user trials. The prototypes showed that PNs need to offer ease of use of ICT services, and that using current off-the-shelf products burdens the users in many ways. Beyond the problem of configuring and setting everything up, which in all trials was done beforehand, other problems encountered included keeping all batteries charged, making sure the communication was up and running, and needing to carry around many devices and their power adapters. On the other hand, users saw the advantage of having PN technologies that can take care of the setup, communication, content distribution, and more.

11.2.2 Final PNP2008 Prototypes

In the final year of the Freeband PNP2008 project, it was decided to steer away from PN application prototyping and to focus the prototyping toward PN-enabling technologies. The result was a small set of prototypes that demonstrated the various PN technologies (PNP2008 2008f).

The main prototype was based on TU Delft's prototype, but was extended with personalization, security, NAT traversal, and some demo applications. In addition, two more prototypes were produced that demonstrated PN service discovery and content management, respectively. Figure 11.7 shows the implementation architecture of the final prototype and the functional components implemented.

Personalization. For the final prototype, both personalization and eviction of nodes were implemented. This was based on the concept of the mother duck or imprinting as explained in Section 9.1.1. The implementation was done in Java and used XML-RPC for communication between the mother duck and the duckling. To make the prototype even more interesting, Bluetooth was used for discovery and communication between the two nodes. The means by which a node became a personal node was based on a certification authority (CA) in the mother duck, called the PNCA. The PNCA gave out

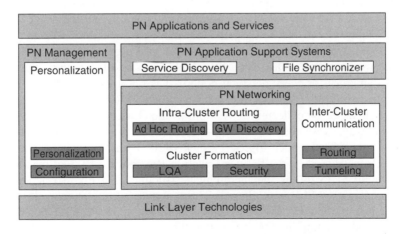

Figure 11.7 PNP2008 final prototype architecture.

X.509 certificates to the personalized nodes that could be used to authenticate other personal nodes and establish secure connections. To avoid man-in-the-middle attacks, PIN numbers were used. Furthermore, the personalization component also supported the uploading of the initial configuration to the newly personalized node.

Cluster formation. Most of this component was based on the TU Delft prototype. However, proper encryption of the intra-cluster traffic was implemented. It was based on IPsec (Kent and Seo 2005) and Internet Key Exchange (IKE) (Kaufman 2005). Using IKE and the certificates handed out by the mother duck during the personalization phase, two personal nodes could authenticate each other on discovering each other's presence. Using a sequence of messages, IKE installed security associations that were used by IPsec to encrypt and decrypt the actual data traffic between the two nodes.

Intra-cluster routing. This component was entirely based on the TU Delft prototype and virtually nothing was modified. It was based on OLSRd (http://www.olsr.org/) with minor modifications to enable the use of improved LQA and support for dynamic gateway nodes.

Inter-cluster communication. The inter-cluster communication component was entirely replaced. The reason for this was that NAT traversal was desired. The prototype was based on a central server through which all inter-cluster traffic had to pass. Using standard virtual private network (VPN) technology, again based on IPsec and IKE, each gateway node connected to the server and thus formed a star topology. The implementation was based on the OpenVPN (http://www.openvpn.net/) software. In this implementation, the PN agent ran an OpenVPN server and each active gateway node ran an OpenVPN client. The clients always tried to connect to the server on the PN agent. Since the OpenVPN tunnels are based on TCP (UDP is also possible) and it was always the clients that initiated the connection, it was possible to traverse almost all NATs that a gateway node might encounter.

The star topology meant that new routing mechanisms were required. The OpenVPN server on the PN agent acts as a layer-two bridge. This meant that no modifications were required to communicate with other gateway nodes in the PN as they will answer the IPv6 neighbor solicitations and the OpenVPN server will bridge accordingly. However, for non-gateway nodes in the clusters, routing functionality was required. To allow for this, each gateway node had to advertise the personal nodes in its cluster to the other gateway nodes. This was implemented using multicast packets via the OpenVPN server to all the other gateway nodes. When receiving such advertisements, a gateway node introduced an entry for every personal node that is not a gateway node. The next hop of that entry would then be the best gateway node to use to communicate with that personal node based on collected LQAs both in the clusters and in the access networks used. In this way, every node could communicate with every other node within the PN.

Service discovery. The service discovery in PNP2008 was implemented as a separate prototype. It was based on the Service Location Protocol (SLP) (Guttman et al. 1999), but was modified to follow the hierarchical structure of a PN. To achieve this, the concept of scope in SLP was altered. Three scopes were defined: cluster, pn, and world. Each scope would have a dedicated directory agent (DA) responsible for that scope. The node running the DA was called the service directory node (SDN). A node with a service or client would, via its SLP service agent or user agent, contact the DA for its cluster (the SDN) using an intra-cluster multicast address. To achieve PN-wide or world-wide service discovery, the DA on the SDN could relay the advertisements and the requests to the DAs responsible for the PN and the world scope.

Most parts of the implementation were based on OpenSLP code (http://www. openslp.org/). However, the DA on the SDN was based on the SLP daemon from the OpenSolaris project (http://www.opensolaris.org/).

File synchronizer. The main goal of the file synchronizer was to implement a content management system for PNs, to allow users to access content on any node from any node within a PN. Also the file synchronizer was implemented as a separate prototype.

The file synchronizer kept track of all the content in the PN and made all content both browsable and searchable from any personal node. When access to a file was requested, it copied the file from the source to the node where it was requested.

For this to work, a server somewhere within the PN was required. That server ran a web server with Web-based Distributed Authoring and Versioning (WebDAV) support. The server kept meta-information about all files in the PN and perhaps copies of some of the files themselves. Every personal node with content had a client that scanned the node for content. When there was a change, the client updated the meta-information on the server using WebDAV. At the same time, any personal node could browse or search the server for content using the collected meta-information. To transfer the actual file content itself, a special agent based on HTTP was used.

To demonstrate the usefulness of the three final PNP2008 prototypes, a handful of demo applications were developed. For the personalization and networking prototype, three demo applications were shown. They included watching a remote video camera and a remote temperature sensor during node and cluster mobility. Furthermore, an indoor location system was demonstrated, where one of the personal nodes could be tracked on a floorplan.

To demonstrate service discovery, a life-sign detection application was made. This application demonstrated the use of a wireless sensor network (WSN) in an environment to detect life-signs. It can be used in home care for the elderly as a sort of warning to the caretakers that something might have happened and assistance might need to be given. The WSN was connected to the PN using a special gateway device. The gateway device is a node in the PN and offered a life-sign detection service to the rest of the PN. Another node in the PN or an approved foreign node may have discovered and connected to the service and accessed the latest activity information.

For the file synchronizer, medical photo file access was shown. The demo also showed that files could be shared with other selected people.

11.3   The MAGNET Prototype

The MAGNET prototype implemented many of the concepts covered in this book. In addition to complete networking and security support, a working service discovery solution, a context management framework, and PNF were provided, as well as some applications demonstrating the benefits of these technologies. After introducing the platforms used in the MAGNET prototype, we cover each of the implemented features one by one.

We do not focus on the MAGNET wireless hardware technologies in this section, but only the software that implemented the concepts mentioned earlier in this book

(Hoebeke et al. 2006b; Louati and Zeghlache 2005; MAGNET Beyond 2008b). The reader that is interested in the MAGNET-developed wireless communication technologies can instead read (MAGNET 2005f).

11.3.1 Hardware and Software Platform

The MAGNET prototype was also based on Linux and involved many different partners. To simplify the integration of the independently developed parts, Ubuntu Linux 7.04 (http://www.ubuntu.com/) was used as the standard operating system. Large parts of the software were written in C or Java; some minor parts were written in C++ or Python. The prototype was designed for using WLAN and Bluetooth as well as the new wireless technologies developed by the project.

The prototype was designed for running on laptops using Ubuntu, but also on the Nokia 770 Internet tablets. The Nokia 770 is a wireless Internet appliance specifically designed for wireless Internet browsing and e-mailing. However, it came with more software, such as Internet radio, media players, and document viewers. The Nokia 770 runs a slimmed-down version of Linux 2.4 for embedded devices called Maemo. The graphical user interface is based on X Windows and GNOME/GTK+. Toward the end of the project, an upgrade to the Nokia N800 was made, which is a successor of the Nokia 770.

Many of the user interfaces of the MAGNET prototype were written for Maemo and the Nokia 770/800. Figure 11.8 shows a screenshot from a Nokia 770 running the MAGNET cluster prototype. It shows how a personal node (Kimmo's_PDA) detects a neighbor node (kisaah-laptop-at-vtt). However, since the neighbor fails to authenticate as a personal node, it is shown as a foreign node with a paler color. Figure 11.9 shows the inter-cluster tunneling application running on a laptop with Ubuntu. It shows how this gateway node

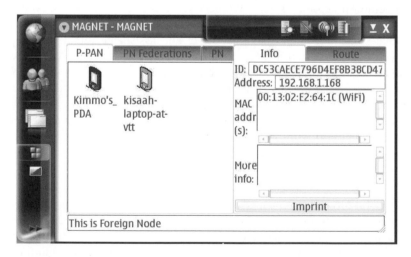

Figure 11.8 Screenshot from the MAGNET cluster formation application. (Reproduced by permission of Kimmo Ahold, © VTT.)

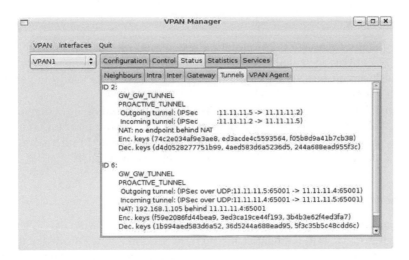

Figure 11.9 Screenshot from the MAGNET inter-cluster tunneling application. (Reproduced by permission of Jeroen Hoebeke, © Ghent University.)

has two active tunnels (ID 2 and ID 6) using IPsec over UDP. One of the tunnel endpoints in the second tunnel is behind an NAT.

11.3.2 PN Networking

The networking parts of the MAGNET prototype were based on the same concepts as the TU Delft and PNP2008 prototypes. However, the implementation was done differently and based on other software and protocols. The networking prototype was built around the universal convergence layer (UCL) (MAGNET 2005d; Sanchez et al. 2005), which was a 2.5-layer implementation that sat in between the network layer and the link layers in the OSI protocol stack, that is, at the same location as the ppand in the TU Delft prototype. The purpose of the UCL was to enable interaction with the underlying link layer technologies in a transparent way and at the same time enable cross-layer optimization. Furthermore, the UCL in MAGNET was extended to also be an enabler for providing link layer security mechanisms, such as ensuring data confidentiality, integrity, authenticity and non-repudiation for intra-PN communication.

The MAGNET prototype also relied much more on kernel modules for the networking implementation instead of using the virtual interface approach. Both neighbor discovery and the UCL were implemented in the kernel for increased performance.

Another big difference was the network layer implementation. The MAGNET prototype was based on IPv4 and used either Wireless Routing Protocol (WRP) (Murthy and Garcia-Luna-Aceves 1996) or Ad Hoc On-Demand Distance Vector (AODV) (Perkins et al. 2003) as the routing protocol. Furthermore, the MAGNET project also experimented with the edge router (ER) concept (see Section 6.4.2), which meant that the gateway node functionality was divided in two parts, one on the gateway node itself and one on the ER. However, the final MAGNET prototype supported scenarios both with and without ERs.

11.3.3 Security

The security of the MAGNET PN prototype was based on the Certified PN Formation Protocol (CPFP) (Mirzadeh et al. 2008a); see Section 9.1.3. It implemented the imprinting procedure and was in turn based on certificates and a PNCA to hand out certificates to the personal nodes.

The communication traffic was protected by encryption based on keys derived using the certificates. The actual encryption was done by the UCL for intra-cluster communication and using dynamic IPsec tunnels for inter-cluster communication. Furthermore, encryption of PN-to-PN traffic in PNFs was also supported using IPsec tunnels.

11.3.4 Service Discovery

Service discovery in the MAGNET prototype was done by the MAGNET Service Management Platform (MSMP) (Ghader et al. 2005). In Section 8.3, we discussed the concepts used in MSMP. MSMP provides context-aware service discovery for PNs and PNFs. It was designed in a modular way and could be extended to support many different service discovery protocols. However, only one protocol was fully implemented; Universal Plug and Play (UPnP). Figure 11.10 shows the various components that make up MSMP.

At the bottom part of Figure 11.10, we can see how MSMP handled the different service discovery protocols using a module for each protocol and an adaptation layer. The protocol-specific modules, such as the Modified UPnP Control Point and Device modules, were mainly responsible for the relaying and translation of service descriptions and operations, such as discovery, registration, and control. At the top of the figure, general functionalities were implemented for common tasks among all service discovery protocols, such as service ranking and interaction with the context management framework and security functionalities.

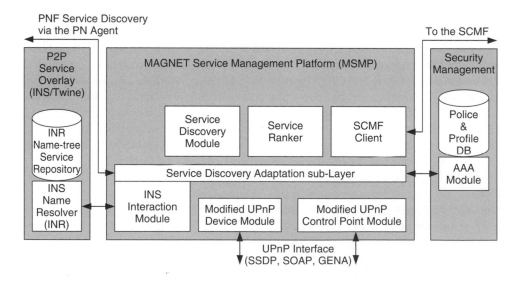

Figure 11.10   The MAGNET Service Management Platform (MAGNET Beyond 2008b).

The peer-to-peer communication between SMNs was based on the INS/Twine framework (Balazinska et al. 2002). Each SMN was a peer in the peer-to-peer overlay that was implemented using a distributed hash table (DHT)-based super-peer overlay network of intentional name resolvers (INRs). Every SMN had an INR and translated service and application information from the supported service discovery protocols into names that could be looked up in the DHT.

11.3.5 Context Management

The MAGNET prototype has support for context management via its Secure Context Management Framework (SCMF) (Sanchez et al. 2006), which we introduced in Section 8.4. The purpose was to have a modular system in which context sources can easily be added and applications can easily obtain the context information collected. The task of the SCMF was to store, process, and deliver context information from the context sources to the context-aware applications in the PN.

Figure 11.11 shows the prototype implementation of the context agent. At the bottom are the data sources. A retriever in the context agent was responsible for obtaining the context information from the source and translating it into the context representation used by SCMF. At the top, context-aware applications connect to the context agent.

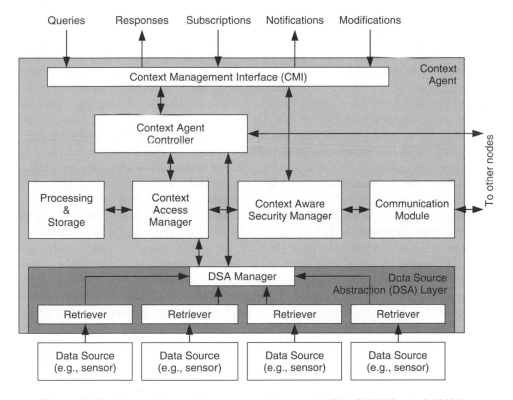

Figure 11.11 Secure Context Management Framework (MAGNET Beyond 2008b).

The applications used a specific query language, which was handled by the Context Management Interface (CMI).

The context requests were handed over to the context aware security manager, which may enforce certain privacy and security policies before passing them on to the context access manager. The latter knew where the context information was located. If it was locally on the node, the request was passed to the data source abstraction layer and the correct retriever. If it was stored or had been processed, the request was sent to the processing and storage module, or if the source was located on another node, the request would be handed over to communication module for forwarding to the correct node.

To support the distribution of context information between personal nodes, the SCMF made use of a concept similar to MSMP: an elected context management node (CMN) for each cluster. The CMN would either collect only location pointers to where context information was available and redirect requests to the node with the source or, alternatively, collect all context information itself. It is the context agent controller shown in Figure 11.11 that is responsible for these tasks.

11.3.6 PN Federations

Perhaps the biggest difference between the MAGNET prototype and the other two prototypes was the fairly complete implementation of PNFs in the MAGNET prototype. It supported the creation of PNFs, participation and maintenance of PNFs, as well as service discovery and sharing in PNFs.

A PN that creates a PNF is called the creator and must generate a PNF profile containing the main details of the PNF and store it in its SCMF. The parts of the PNF profile that describe the PNF were sent to a central directory, called the PN Directory Server (PNDS). Other PNs could query the PNDS to find available PNFs. A PN that wished to join an existing PNF may go into a dialog with the creator to see whether it was allowed to enter the PNF or not. The PNDS also acted as a CA. It authenticated the PNs and used the CPFP to establish the necessary security associations to be used for the PNF communication.

To enable communication between the PNs in a PNF, the MAGNET prototype relied on network overlays. Each PNF had its own network overlay with its own address space. The network overlay was based on the same solutions as PN networking. Hence, the members were interconnected using dynamic tunneling similar to the PN formation.

To support sharing and discovering of services in a PNF, a service overlay was implemented. It was based on the PN-level MSMP and also used INS/Twine to establish a peer-to-peer service overlay for each PNF. In each PNF member, a new component, named the federation agent, was introduced. The federation agent acted as super-peer within the service overlay in the same way as a SMN was a peer in the PN-wide service discovery solution. In a similar way, each member used its federation agent to publish and discover services in the PNF by translating service descriptions into intentional names for insertion into the DHT of the PNF.

11.3.7 Applications

Several PN applications were developed to various degrees in MAGNET. Two applications were developed to such a degree that they could be tested on top of the MAGNET prototype with real users. Those two applications were the Icebreaker and the Life Style Companion (MAGNET Beyond 2008c).

The aim of the Icebreaker application was to demonstrate the PN capabilities with regard to socializing at a big event, such as a conference. Conference delegates would each have a PN and be able to use an 'icebreaker service' offered by the conference organizers. The service could be accessed from a Nokia N800 and the users could use it to subscribe and check in to the conference. It also provided a matching service, in which delegates upload some parts of their virtual business card, such as name, contact information, and (business) interests. The service would then match the delegates and suggest that they meet other delegates present, based on common interests. The service could also be used to access equipment in a showroom to display presentations.

The Life Style Companion application was an electronic personal trainer for the PN user. When the user entered a gym, the application activated and asked the user to first step onto a scale. The weight was taken and then the application instructed the user to use a certain piece of fitness equipment. In the case of the MAGNET pilot, it was an exercise bike. The application would detect the user on a bike and configure it according the its exercise program.

Both pilot prototypes were implemented using the PN technologies developed by the MAGNET projects. For instance, the virtual business cards and the scale both relied on the SCMF. Using PN networking, the delegate could access a file stored on a remote device and present it in the showroom. PNFs were used to connect the delegates' PNs with a special PN established by the event to enable access to the event services. In the gym scenario, a PNF was formed between the user's PN and the gym's PN. The bike was offered as a service in that PNF via MSMP. The personal trainer application, which was running on the user's PN device (a Nokia N800), accessed the bike service to configure the load, etc. In the end, most parts of the MAGNET prototype developed were actually used in the pilot tests.

Despite some issues, such as the slow computing capabilities of the Nokia N800 devices (and the non-optimized prototype software), which also led to stability issues, the pilot tests were successful and much was learned (MAGNET Beyond 2008c). In general, the users were positive toward the PN concepts and technologies and demonstrated a clear interest in using them in their everyday life. However, some concepts were new to them and required some explanation before all understood the ideas. The learning period was short though, and after experiencing the system, the users could understand and use most concepts without difficulty. The users also showed interest in PNFs, which were seen as a good way of organizing people with whom they would like to share tools and data.

Hence, it was concluded that the PN concepts and technologies have great potential. Most users found the concepts and services interesting and useful. Hence, we expect increased demand for PN technologies in the future.

11.4 Summary

A very important way of exploring the PN concepts is via prototyping. In this chapter, we have outlined three different prototypes. We started with the TU Delft prototype, which mainly consisted of network-related functionalities. This prototype demonstrated how the network concepts that have been proposed in this book can be implemented in a concrete way.

In the subsequent sections, we introduced the prototypes developed by PNP2008 and MAGNET. Details were given of the implementation of various functionalities, such as security, content distribution, service discovery, context management, and PNF. The PN applications implemented were briefly introduced, as well as highlights from the MAGNET user trials. It was concluded that most users found the PN concepts and services promising and useful.

12

The Future of Personal Networks

Today, wireless communication technologies are specialized toward particular communication problems in order to better address certain niche problems. This causes difficulties for end-users who have to understand and master all these technologies and accept that they do not fully work together. Hence, the main problem is how to make these technologies best complement each other and work together seamlessly. There are plenty of partial solutions that can solve individual problems, but no overall solution; and the individual partial solutions are frequently incompatible with each other. There is a clear lack of concertation among technologies to reach the vision of personal networks.

In this book, we have introduced the concept of personal networks, a future concept of advanced personal communication. We started out by showing that there is a real benefit in having seamless communication among all one's electronic devices and that this can lead to new applications and services. It will be easier for service providers to offer added-value services to a person if there is a personal communication platform to build on. Hence, the concept of personal networks is an attractive proposition with many applications.

12.1 Are We There Yet?

There are already several prototypes that together demonstrate most aspects of personal networks. However, these are just prototypes, not ready products that end-users can acquire and start to use. Hence, personal networks are only a reality in research laboratories. But how far have we come with the personal network technology? Is it ready for large scale deployment?

To answer that question, it is necessary to go back to the requirements – not only the high level architectural requirements in Chapter 2, but also all the requirements concerning the individual components of the architecture given in other chapters. Regarding the requirements on the architecture, several issues still remain, as discussed in Chapter 4. Such issues include whether PNs will be able to meet the demands of security, whether the PNs actually will be as easy to use as we hope, and whether PNs can support the social interactions among users. Unfortunately, we believe there is only one way to find out, and that is to build a prototype and test it with real users.

Personal Networks: Wireless Networking for Personal Devices   Martin Jacobsson, Ignas Niemegeers and Sonia Heemstra de Groot
© 2010 John Wiley & Sons, Ltd

Elsewhere in this book, we identified the following technical gaps:

 (i) We stressed the need for support of heterogeneous link layer technologies, but this
      has rarely been studied. Hence, more research is required regarding PNs with mul-
      tiple simultaneous link layer technologies, especially in the area of routing.
 (ii) The multi-hop routing in clusters is far from perfect. We believe the routing quality
      is satisfactory for the majority of applications. However, other requirements, such
      as low overhead and low power consumption, have more or less been neglected.
(iii) Routing over infrastructure networks is an issue. The approach taken in Chapter 6 is
      to design a system that works over as many connection types as possible. However,
      many systems are not built for PN communication and will cause problems, such as
      during handovers. Furthermore, there is a lack of cooperation between the various
      alternatives, which may hamper handovers between technologies.
 (iv) We indicated the lack of intelligent middleware that observes and understands the
      underlying layers, and can take actions based on that as well as on behalf of the user
      and the applications. Hence, techniques such as service discovery and management,
      context management, and cognition must significantly improve.
 (v) Network layer anonymity and privacy is not what it should be. In a world where
      a person's devices communicate with each other, privacy becomes a greater issue.
      Anonymity is one of the few solutions to this problem. However, we have only
      been able to identify anonymity solutions for intra-PN communication, but not for
      the general inter-PN communication case. Furthermore, most link layer technologies
      need to be enhanced with better anonymity features.
 (vi) Protection against denial of service attacks, viruses, and other threats. How can a
      PN handle security and trust in a way that is easy to use?
(vii) We have defined some solutions for personal network federations. However, to
      make PNFs happen, more work is required in flexible authentication mechanisms,
      reputation management, and service discovery.

However, the biggest obstacle to the widespread usage of PNs is the current lack
of PN applications, PN-capable devices, and standardization. The real success of PNs
will only be achieved when most device manufacturers, network providers, and content
providers offer PN-enabled products and services. However, that requires far-reaching
interoperability.

12.2   Future Directions

To achieve wide PN deployment, all players, from different parts of the industry, must
start a collaborative effort, and that requires good incentives for each of them. PNs must
be beneficial not only for the user, but also for manufacturers, network providers, service
providers, etc.

For network providers, benefits will come from increased communication needs in
terms of greater bandwidth requirements, the need to be constantly connected, and to
be connected through more types of connections. Network providers that can offer not
only one type of access, but a multitude of infrastructure-based connection types, such as

UMTS, WiMAX, WLAN, Cable, and DSL, will have an advantage if they can provide one single seamless and well-coordinated network service.

Content and service providers will mainly benefit from a unified platform through which they can offer their content and services to many more users in a homogeneous and secure way. The mobility aspects of personal networks will make it possible for the providers to almost always reach their customers with their services.

Device manufacturers will probably face the biggest challenges, since industries as diverse as manufacturers of consumer electronics, mobile products, PC equipment, home appliances, home automation, and many more need to be involved. In the short term, manufacturers that can offer devices capable of participating in people's PNs will have a leading edge. In the longer term, we believe that PNs will spur new sorts of applications requiring new types of electronic devices.

The work must now also start as an industrial effort. Standardization, cross-industry liaisons, and alliances that can safeguard a smooth and interoperable development of PNs are necessary. It is certainly a positive sign for PNs that standardization and other business-related efforts are starting to be discussed. Examples include one PN standardization group within Ecma International (http://www.ecma-international.org/memento/TC32-PNF-M.htm) and one group dealing with health-related PNs within ETSI (ETSI 2009). This, together with an increased awareness and demand from users, will make PNs happen.

Appendix A

Terminology

In this appendix, we define the terminology used in the book. This will be useful for readers who wish to get a more in-depth and precise understanding of the terms and concepts. The definitions can also serve the standardization of PN concepts.

The terms and concepts in this appendix are similar to those defined in the IST MAGNET and MAGNET Beyond projects (Jacobsson et al. 2005b; MAGNET 2005e; Muñoz et al. 2005) and the PNP2008 project (PNP2008 2008e), all of which are evolutions of Jacobsson et al. (2004) and Jacobsson and Niemegeers (2005).

We introduce the terms and their definitions based on the abstraction level they belong to. In the definitions, we use italics when we refer to a defined term. Terms related to personal network federations are listed separated in the last section.

A.1 Connectivity Abstraction Level

Device: A physical entity that aggregates physical capabilities, such as processing capabilities, communication capabilities, memory storage, display, loudspeakers, keyboard, etc., at a single physical location.

Communication interface: A module that can send and receive data packets according to a particular physical layer technology and medium access control mechanism.

Communication domain: A collection of *communication interfaces* using a common communication technology that are controlled by a single medium access control mechanism (either centralized, distributed, or a combination of both).

A.2 Network Abstraction Level

PN-capable device: A *device* that can house at least one *node*. This means that the *device* can fully participate in the PN networking mechanisms. A *PN-capable device* that can house more than one *node* and thereby be shared by several PNs is sometimes also referred to as a *multi-PN-capable device*.

Node: A communicating entity or network termination point residing on a *PN-capable device*. A *node* must implement the protocols and mechanisms required to form PNs. Note that for most *devices* that have one exclusive user, a *node* is practically equal to

Personal Networks: Wireless Networking for Personal Devices   Martin Jacobsson, Ignas Niemegeers and Sonia Heemstra de Groot
© 2010 John Wiley & Sons, Ltd

a *device*. When a *device* has more than one user, it may have more nodes on it, one for each user.

Personal node: A *node* related to a given person (e.g. ownership). Two *nodes* related to the same person are personal to each other and have a permanent security association with each other.

Personal network (PN): The set of all the *nodes* related to a given *PN* user. Each *personal node* in a *PN* has a permanent security association to all the other *Personal Nodes* in the *PN*.

Cluster: A connected network of *personal nodes* using only the *personal nodes* and the *communication domains* they are part of. This means that two *personal nodes* are in the same *cluster* if they can communicate using a path between them consisting of only *personal nodes* and their *communication domains*. A single *personal node* with no other *personal nodes* in its communication range is by itself a *cluster*.

Foreign node: A *node* that has a different trust attribute and therefore is not part of the (same) PN. *Foreign nodes* can either be trusted or non-trusted. Whenever trusted, they will typically have an ephemeral security association with one or more *nodes* in the PN.

Interconnecting structures: Public, private, or shared wired, wireless, or hybrid networks such as an UMTS network, the Internet, an intranet, or an ad hoc network that can be used to interconnect *clusters*. We also assume that all *interconnecting structures* are connected to each other.

Gateway node: A *personal node* within a *cluster* that enables connectivity to *foreign nodes* and non-PN-enabled *devices* outside the *cluster*. Some *gateway nodes* have access to *interconnecting structures* and can then connect to other distant personal *clusters* via the *gateway nodes* in those *clusters* by means of inter-*cluster* tunnels.

PN agent: The *PN agent* is an infrastructure-based logical entity accessible through the *interconnecting structure*. Its task is to keep track of each *cluster* and their attachment points to the *interconnecting structure* so that it can assist in the establishment and maintenance of inter-*cluster* tunnels. It may also be an entry point for *foreign nodes* that want to use *services* offered by the PN. Each PN should have a *PN agent*.

A.3 Application and Service Abstraction Level

Application: A logical component implemented by a software program running on a *node* that may be started as a direct consequence of an action of a *PN* user, for example, the *PN* user selects the application from a menu and presses a 'Run' button or something similar.

Service: A logical component implemented by a software program running on a *node* that offers something that an *application* can use through a specified interface. *Services* have formalized descriptions that enable *applications* to discover useful *services*. The description must describe what the *service* offers, how to interface with it, and on which *node* it is offered.

Client: An *application* (or *service*) that can use *services*. A *service* discovery framework must be able to find the *services* that *clients* require. *Clients* can run on both *personal nodes* and *foreign nodes*.

Context: Any information that can be used to characterize the situation of a *service* relevant to the interaction between a user and an *application*, including the user and *application*s themselves. Some typical examples of context are location, weather, time, and the user's mood.

Content: Information and experiences that provide value for a *PN* user. Content may take the form of a *service* – text (e.g. documents), multimedia files (e.g. audio or video files) or any other file type which follows a *content* life cycle and requires management (e.g. executables).

Service proxy: A *service*, running on a *gateway node*, which can act as both *client* and *service* for one or more types of *service*s. Its main task is to connect *client*s and *service*s within the *PN* with *client*s and *service*s outside the *PN* without requiring end-to-end network connections. It constitutes one way of doing foreign communication.

A.4   Personal Network Federations

PN federation (PNF): A temporal, ad hoc, purpose- or opportunity-driven collaboration of two or more PNs.

Federation manager: Functionality to create and manage the *PN federation*.

Federation agent: Functionality within a *PN*, responsible in *PN federation*-related activities, such as joining/leaving a *PN federation* and controlling the access to *PN* resources and *service*s.

PNF profile: A set of one or more distinguishing characteristics of a *PN federation*, such as its goal, name and/or identity, contact information, and *service* list.

PNF participation profile: A set of one or more characteristics of a *PN* willing to federate, such as credentials and list *service*s.

PNF member: A *PN* that is authorized to be a participant of a particular *PN federation*.

Membership credential: A special token or credential granted to the authorized *PN federation* member.

References

3GPP (2005) Specification of the subscriber identity module – mobile equipment (SIM-ME) interface. Technical Specification, 3GPP, TS 51.011, Version 4.15.

3GPP (2009a) Personal network management (PNM) – Procedures and information flows – Stage 2. Technical Specification, 3GPP TS 23.259V9.1.0 (2009-09).

3GPP (2009b) Service requirements for personal network management (PNM) – Stage 1. Technical Specification, 3GPP TS 22.259V9.1.0 (2009-09).

Aarts, E. and Encarnaçào, J. (2006) *True Visions: The Emergence of Ambient Intelligence*. Springer-Verlag, Berlin.

Abbate, J. (1999) *Inventing the Internet*. MIT Press, Cambridge, MA.

Abolhasan, M., Wysocki, T. and Dutkiewicz, E. (2004) A review of routing protocols for mobile ad hoc networks. *Ad Hoc Networks*, **2**(1), 1–22.

Abowd, G.D., Dey, A.K., Brown, P.J., Davies, N., Smith, M. and Steggles, P. (1999) Towards a better understanding of context and context-awareness. First International Symposium on Handheld and Ubiquitous Computing (HUC'99), Karlsruhe, Germany.

Aguayo, D., Bicket, J., Biswas, S., Judd, G. and Morris, R. (2004) Link-level measurements from an 802.11b mesh network. ACM SIGCOMM Conference 2004, Portland, OR, USA.

Almesberger, W. (1999) Linux network traffic control – implementation overview. Fifth Annual Linux Expo, Raleigh, NC, USA.

Alutoin, M., Lehtonen, S., Ahola, K. and Paananen, J. (2007) Personal network directory service. *Telektronikk* **1.07**, 85–92.

Anastasi, G., Borgia, E., Conti, M. and Gregori, E. (2004) Wi-Fi in ad hoc mode: a measurement study. Second IEEE Annual Conference on Pervasive Computing and Communications (PerCom'04), Orlando, FL, USA.

Ansari, F. and Sathyanath, A. (2007) STEM: Seamless transport endpoint mobility. *ACM SIGMOBILE Mobile Computing and Communications Review (MC2R)*, **11**(2), 1–13.

Arvind and Hicks, J. (2006) A mobile phone ecosystem: MIT and Nokia's joint research venture. *IEEE Intelligent Systems*, **21**(5), 78–79.

Awerbuch, B., Holmer, D. and Rubens, H. (2003) High throughput route selection in multi-rate ad hoc wireless networks. Technical Report Version 2, John Hopkins University, Baltimore, MD, USA.

Balazinska, M., Balakrishnan, H. and Karger, D. (2002) INS/Twine: A scalable peer-to-peer architecture for intentional resource discovery. First International Conference on Pervasive Computing, Zurich, Switzerland.

Balfanz, D., Durfee, G., Smetters, D.K. and Grinter, R.E. (2004) In search of usable security: Five lessons from the field. *IEEE Security and Privacy*, **2**(5), 19–24.

Balfanz, D., Smetters, D., Stewart, P. and Wong, C. (2002) Talking to strangers: Authentication in ad-hoc wireless networks Network and Distributed System Security Symposium (NDSS'02), San Diego, CA, USA.

Bicket, J.C., Aguayo, D., Biswas, S. and Morris, R. (2005) Architecture and evaluation of an unplanned 802.11b mesh network. Annual International Conference on Mobile Computing and Networking (MobiCom'05), Cologne, Germany.

Braun, K., Grollman, J., Horn, M., Raffler, H., Thulke, W. and Weigel, W. (1993) Universal personal networking. Second International Conference on Universal Personal Communications (ICUPC'93), Ottawa, Canada.

Brik, V., Rayanchu, S., Saha, S., Sen, S., Shrivastava, V. and Banerjee, S. (2008) A measurement study of a commercial-grade urban WiFi mesh. Eighth ACM SIGCOMM Conference on Internet Measurement (IMC'08), Vouliagmeni, Greece.

Broch, J., Maltz, D.A., Johnson, D.B., Hu, Y.C. and Jetcheva, J. (1998) A performance comparison of multihop wireless ad hoc network routing protocols. Fourth Annual ACM/IEEE International Conference on Mobile Computing and Networking (MOBICOM'98), Dallas, TX, USA.

Buyya, R. and Venugopal, S. (2005) A gentle introduction to grid computing and technologies. *CSI Communications*. http://www.buyya.com/papers/GridIntro-CSI2005.pdf.

Calhoun, P. and O'Hara, B. (2005) *802.11r strengthens wireless voice*. Network World. http://www.networkworld.com/news/tech/2005/082205techupdate.html.

Carter, C., Kravets, R. and Thourrilhes, J. (2003) Contact Networking: A localized mobility system. First International Conference on Mobile Systems, Applications, and Services (MobiSys'03), San Francisco.

CempakaWangi, N.I., Prasad, R.V., Jacobsson, M. and Niemegeers, I.G. (2008) Address autoconfiguration in wireless ad hoc networks: Protocols and techniques. *IEEE Wireless Communications Magazine*, **15**(1), 70–80.

Chakeres, I.D. and Belding-Royer, E.M. (2002) The utility of hello messages for determining link connectivity. Fifth International Symposium on Wireless Personal Multimedia Communications, Honolulu, HI, USA.

Chakeres, I.D. and Perkins, C.E. (2009) Dynamic MANET on-demand (DYMO) routing. IETF Internet-Draft (Work in Progress), draft-ietf-manet-dymo-17.

Chelius, G. and Fleury, E. (2002) Ananas: A local area ad hoc network architectural scheme. Fourth International Workshop on Mobile and Wireless Communications Network (MWCN'02), Stockholm, Sweden.

Choi, J. and Daley, G. (2005) Goals of detecting network attachment in IPv6. IETF RFC 4135.

Cisco Systems (2009) Scaling the mobile Internet, White Paper. http://www.cisco.com/en/US/solutions/collateral/ns341/ns523/white_paper_c11-523350.html.

Clausen, T.H. and Jacquet, P. (2003) Optimized link state routing protocol (OLSR). IETF RFC 3626.

Clausen, T.H., Dearlove, C.M. and Dean, J.W. (2009a) MANET neighborhood discovery protocol (NHDP). IETF Internet-Draft (Work in Progress), draft-ietf-manet-nhdp-11.

Clausen, T.H., Dearlove, C.M. and Jacquet, P. (2009b) The optimized link state routing protocol version 2. IETF Internet-Draft (Work in Progress), draft-ietf-manet-olsrv2-10.

Clausen, T.H., Dearlove, C.M., Dean, J.W. and Adjih, C. (2009c) Generalized MANET packet/message format. IETF RFC 5444.

David, K. (2008) *Technologies for the Wireless Future: Wireless World Research Forum (WWRF), Volume 3*. John Wiley & Sons, Ltd, Chichester.

De Couto, D.S.J., Aguayo, D., Bicket, J. and Morris, R. (2003a) A high-throughput path metric for multi-hop wireless routing. Annual International Conference on Mobile Computing and Networking (MobiCom'03), San Diego, CA, USA.

De Couto, D.S.J., Aguayo, D., Chambers, B.A. and Morris, R. (2002) Effects of loss rate on ad hoc wireless routing. Technical Report MIT-LCS-TR-836, Massachusetts Institute of Technology, USA

De Couto, D.S.J., Aguayo, D., Chambers, B.A. and Morris, R. (2003b) Performance of multihop wireless networks: Shortest path is not enough. *ACM SIGCOMM Computer Communications Review*, USA.

Debaty, P. and Caswell, D. (2001) Uniform web presence architecture for people, places, and things. *IEEE Personal Communications*, **8**(4), 46–51.

Deering, S.E. and Hinden, R.M. (1998) Internet protocol, version 6 (IPv6) specification. IETF RFC 2460.

den Hartog, F., Blom, M., Lageweg, C., Peeters, M., Schmidt, J., van der Veer, R., de Vries, A., van der Werff, M., Tao, Q. and Veldhuis, R. (2007) First experiences with personal networks as an enabling platform for service providers. Second International Workshop on Personalized Networks (Pernets'07), Philadelphia, PA, USA.

Devarapalli, V., Wakikawa, R., Petrescu, A. and Thubert, P. (2005) Network mobility (NEMO) basic support protocol. IETF RFC 3963.

Dey, A.K. (2000) Providing architectural support for building context-aware applications. PhD thesis College of Computing, Georgia Institute of Technology, USA.

Dierks, T. and Rescorla, E. (2008) The transport layer security (TLS) protocol – Version 1.2. IETF RFC 5246.

Dodson, S. (2003) The Internet of things, *The Guardian*. http://www.guardian.co.uk/technology/2003/oct/09/ shopping.newmedia.

Dornan, A. (2001) *The Essential Guide to Wireless Communications Applications*. Prentice Hall, Upper Saddle River, NJ.

Dourish, P., Grinter, R.E., Delgado de la Flor, J. and Joseph, M. (2004) Security in the wild: User strategies for managing security as an everyday practical problem. *Personal and Ubiquitous Computing*, **8**(6), 391–401.

Draves, R., Padhye, J. and Zill, B. (2004) Routing in multi-radio, multi-hop wireless mesh networks. Annual International Conference on Mobile Computing and Networking (MobiCom'04), Philadelphia, PA, USA.

Droms, R. (1997) Dynamic host configuration protocol. IETF RFC 2131.

Dunlop, J. (2004) The concept of a personal distributed environment for wireless service delivery. NEXWAY White Paper. http://www.telecom.ntua.gr/nexway/php/index.php?action=downloadfile&filename=pde_white_paper.pdf&directory=public_downloads/WHITE%20PAPERS&.

Dunlop, J., Atkinson, R., Irvine, J.M. and Pearce, D. (2003) A personal distributed environment for future mobile systems. 12th IST Mobile & Wireless Communication Summit, Aveiro, Portugal.

ETSI (2009) eHealth architecture: User service models and application classification into service models. TR 102 764. http://webapp.etsi.org/workProgram/Report_WorkItem.asp?wki_id=27268.

Ford, B. (2003) Unmanaged Internet Protocol: Taming the edge network management crisis. Second Workshop on Hot Topics in Networks (HotNets-II), Cambridge, MA, USA.

Garfinkel, S. (2000) *Database Nation: The Death of Privacy in the 21st Century*. O'Reilly and Associates, Beijing.

Garg, V. (2007) *Wireless Communications and Networking*. Morgan Kaufmann, San Francisco.

Gehrmann, C., Kuhn, T., Nyberg, K. and Windirsch, P. (2002a) Trust model, communication and configuration security for personal area networks. 11th IST Mobile & Wireless Telecommunications Summit, Thessaloniki, Greece.

Gehrmann, C., Nyberg, K. and Mitchell, C. (2002b) The personal CA – PKI for a personal area network. 11th IST Mobile & Wireless Telecommunications Summit, Thessaloniki, Greece.

Ghader, M., Olsen, R.L., Genet, M.G. and Tafazolli, R. (2005) Service management platform for personal networks. 14th IST Mobile & Wireless Communications Summit, Dresden, Germany.

Ghader, M., Olsen, R.L., Prasad, R.V., Jacobsson, M., Sanchez, L., Lanza, J., Louati, W., Genet, M.G., Zegh-lache, D. and Tafazolli, R. (2006) Service discovery in personal networks: design, implementation and analysis. 15th IST Mobile & Wireless Communications Summit, Mykonos, Greece.

Guttman, E., Perkins, C., Veizades, J. and Day, M. (1999) Service location protocol, version 2. IETF RFC 2608.

Haas, Z.J. and Pearlman, M.R. (2001) The performance of query control schemes for the zone routing protocol. *IEEE/ACM Transactions on Networking*, **9**(4), 427–438.

Hall, J., Barbeau, M. and Kranakis, E. (2003) Detection of transient in radio frequency fingerprinting using signal phase. Wireless and Optical Communications Conference (WOC'03), Banff, AB, Canada.

He, K. (2005) Kernel korner – Why and how to use netlink socket. *Linux Journal*. http://www.linuxjournal.com/article/7356.

Heijenk, G.J. and Liu, F. (2006) Interference-based routing in multi-hop wireless infrastructures. *Computer Communications*, **29**(13–14), 2693–2701.

Hinden, R.M. and Deering, S.E. (2006) IP version 6 addressing architecture. IETF RFC 4291.

Hoebeke, J. (2007) Adaptive ad hoc routing and its application to virtual private ad hoc networks. PhD thesis Universiteit Gent, Belgium.

Hoebeke, J., Holderbeke, G., Moerman, I., Jacobsson, M., Prasad, R.V., Wangi, N.I.C., Niemegeers, I.G. and Heemstra de Groot, S.M. (2006a) Personal network federations. 15th IST Mobile & Wireless Communications Summit, Mykonos, Greece.

Hoebeke, J., Holderbeke, G., Moerman, I., Louati, W., Louati, W., Genet, M.G., Zeghlache, D., Sanchez, L., Lanza, J., Alutoin, M., Ahola, K., Lehtonen, S. and Pallares, J.J. (2006b) Personal networks: from concept to a demonstrator. 15th IST Mobile & Wireless Communications Summit, Mykonos, Greece.

Husemann, D., Narayanaswa, C. and Nidd, M. (2004) Personal mobile hub. Eighth IEEE International Symposium on Wearable Computers (ISWC'04), Arlington, VA, USA.

Huttunen, A., Swander, B., Volpe, V., DiBurro, L. and Stenberg, M. (2005) UDP encapsulation of IPsec ESP packets. IETF RFC 3948.

Ibrohimovna, M. and Heemstra de Groot, S.M. (2009) Policy-based hybrid approach to service provisioning in federations of personal networks. Third International Conference on Mobile Ubiquitous Computing, Systems, Services and Technologies (UBICOMM'09), Sliema, Malta.

IEEE (1997) Guidelines for 64-bit global identifier (EUI-64) registration authority. http://standards.ieee.org/regauth/oui/tutorials/EUI64.html.

IEEE (1999) Part 11: Wireless LAN medium access control (MAC) and physical layer (PHY) specifications. ANSI/IEEE Std 802.11, ISO/IEC 8802-11: 1999.

IEEE (2003) Part 15.3: Wireless medium access control (MAC) and physical layer (PHY) specifications for high rate wireless personal area networks (WPAN). IEEE Std 802.15.3 (2003).

IEEE (2004a) Part 11: Wireless LAN medium access control (MAC) and physical layer (PHY) specifications – Amendment 6: Medium access control (MAC) security enhancements. IEEE Std. 802.11i (2004).

IEEE (2004b) Part 16: Air interface for fixed broadband wireless access systems. IEEE Std 802.16 (2004).

IEEE (2005) Part 15.1: Wireless medium access control (MAC) and physical layer (PHY) specifications for wireless personal area networks (WPANs(TM)). ANSI/IEEE Std 802.15.1 (2005).

IEEE (2006a) Amendment 2: Physical and medium access control layers for combined fixed and mobile operation in licensed bands and corrigendum 1. IEEE Std 802.16e-2005 and IEEE Std 802.16-2004/Cor 1-2005.

IEEE (2006b) Part 15.3b: Wireless medium access control (MAC) and physical layer (PHY) specifications for high rate wireless personal area networks (WPANs) Amendment 1: MAC sublayer. IEEE Std. 802.15.3b (2005).

ISO (1996) Information technology – Open systems interconnection – Basic reference model: The basic model. ISO/IEC 7498-1:1994(E).

ISO (2008) Information technology – UPnP device architecture – Part 1: UPnP device architecture version 1.0. ISO/IEC 29341-1:2008.

ISTAG (2001) Scenarios for ambient intelligence in 2010. Technical report, ftp://ftp.cordis.lu/pub/ist/docs/istagscenarios2010.pdf. Information Society Technologies Advisory Group (ISTAG).

ITU (2005) ITU Internet reports 2005: The Internet of things. Technical report, http://www.itu.int/publ/S-POL-IR.IT-2005/e.

Jacobsson, M. and Niemegeers, I.G. (2005) Privacy and anonymity in personal networks Second International Workshop on Pervasive Computing and Communication Security (PerSec'05), Kauai Island, Hawaii, USA.

Jacobsson, M., Guo, C. and Niemegeers, I.G. (2005a) A flooding protocol for MANETs with self-pruning and prioritized retransmissions. International Workshop on Localized Communication and Topology Protocols for Ad hoc Networks (LOCAN'05), Washington DC, USA.

Jacobsson, M., Hoebeke, J., Heemstra de Groot, S.M., Lo, A., Moerman, I. and Niemegeers, I.G. (2004) A network layer architecture for personal networks. First MAGNET Workshop, Shanghai, China.

Jacobsson, M., Hoebeke, J., Heemstra de Groot, S.M., Lo, A., Moerman, I., Niemegeers, I.G., Muñoz, L., Alutoin, M., Louati, W. and Zeghlache, D. (2005b) A network architecture for personal networks. 14th IST Mobile & Wireless Communications Summit, Dresden, Germany.

Jacobsson, M., Prasad, R.V., Lu, W. and Niemegeers, I.G. (2006) Foreign communication in personal networks. Fifth Annual Mediterranean Ad Hoc Networking Workshop (Med-Hoc-Net'06), Lipari, Italy.

Jefferies, N. (2007) Global vision for a wireless world. 18th Wireless World Research Forum (WWRF) Meeting, Helsinki, Finland.

Jehangir, A. and Heemstra de Groot, S.M. (2007) Securing inter-cluster communication in personal networks. Second International Workshop on Personalized Networks (Pernets'07), Philadelphia, PA, USA.

Johnson, D.B. and Maltz, D.A. (1996) Dynamic source routing in ad-hoc wireless network. ACM SIGCOMM Conference 1996, Stanford University, CA, USA.

Johnson, D.B., Maltz, D.A. and Hu, Y.C. (2007) The dynamic source routing protocol (DSR) for mobile ad hoc networks. IETF RFC 4728.

Johnson, D.B., Perkins, C.E. and Arkko, J. (2004) Mobility support in IPv6. IETF RFC 3775.

Jokela, P., Moskowitz, R. and Nikander, P. (2008) Using the encapsulating security payload (ESP) transport format with the host identity protocol (HIP). IETF RFC 5202.

Joshi, J. (2004) Access-control language for multidomain environments. *IEEE Internet Computing*, **8**(6), 40–50.

Kaashoek, F. and Morris, R. (2006) User-relative names for globally connected personal devices. Fifth International Workshop on Peer-to-Peer Systems (IPTPS'06), Santa Barbara, CA, USA.

Kahn, R.E., Gronemeyer, S.A., Burchfiel, J. and Kunzelman, R.C. (1978) Advances in packet radio technology. *Proceedings of the IEEE*, **66**(11), 1468–1496.

Kaufman, C. (2005) Internet key exchange v2. IETF RFC 4306.

Kent, S. (2005) IP encapsulating security payload (ESP). IETF RFC 4303.

Kent, S. and Seo, K. (2005) Security architecture for the Internet protocol. IETF RFC 4301.

Komu, M., Henderson, T., Tschofenig, H., Melen, J. and Keraenen, A. (2009) Basic HIP extensions for traversal of network address translators. IETF Internet-Draft (Work in Progress), draft-ietf-hip-nat-traversal-08.

Koodli, R. (2009) Mobile IPv6 fast handovers. IETF RFC 5568.

Kravets, R., Carter, C. and Magalhães, L. (2001) A cooperative approach to user mobility. *ACM Computer Communications Review*, **31**(5), 57–69.

Krishnan, S. and Daley, G. (2009) Simple prodecures for detecting network attachment in IPv6. IETF Internet-Draft (Work in Progress), draft-ietf-dna-simple-11.

Laganier, J. and Egger, L. (2008) Host identity protocol (HIP) rendezvous extension. IETF RFC 5204.

Laganier, J., Koponen, T. and Eggert, L. (2008) Host identity protocol (HIP) registration extension. IETF RFC 5203.

Lipman, J., Boustead, P. and Chicharo, J. (2004) Reliable optimised flooding in ad hoc networks. IEEE Sixth CAS Symposium on Emerging Technologies: Frontiers of Mobile and Wireless Communication, Shanghai, China.

Louagie, F., Muñoz, L. and Kyriazakos, S. (2003) Paving the way for the fourth generation: A new family of wireless personal area networks. 12th IST Mobile & Wireless Communications Summit, Aveiro, Portugal.

Louati, W. and Zeghlache, D. (2005) Network-based virtual personal overlay networks using programmable virtual routers. *IEEE Communications Magazine*, **43**(8), 86–94.

Macker, J. (2009) Simplified multicast forwarding for MANET. IETF Internet-Draft (Work in Progress), draft-ietf-manet-smf-09.

MAGNET (2004a) Architectures and protocols for ad-hoc self-configuration, interworking, routing and mobility. Deliverable IST-507102 MAGNET/WP2.4/IMEC/D2.4.1/PU/001/1.0.

MAGNET (2004b) Resource and service discovery: PN solutions. Deliverable IST-507102 MAGNET/WP2.2/UNIS/D2.2.1/R/PU/001/1.0.

MAGNET (2005a) Ad-hoc self organising and routing architectures (NETWORK layer). Deliverable IST-507102 MAGNET/WP2.3/RWTH/D2.3.2/PU/001/19.12.2005.

MAGNET (2005b) Final user requirements for the PN service architecture. Deliverable IST-507102 MAGNET/WP1.1/DTU/D1.1.1c/R/PU/001/20.12.2005.

MAGNET (2005c) Final version of the network-level security architecture specification. Deliverable IST-507102 MAGNET/WP4.3/UNIS/D4.3.2/PU/1.00.

MAGNET (2005d) MAC/RRM schemes for WPAN (Update D3.3.2a). Deliverable IST-507102 MAGNET/WP3.3/UNIS/D3.3.2b/R/PU/001/1.1.

MAGNET (2005e) Overall secure PN architecture. Deliverable IST-507102 MAGNET/WP2.1/RWTH/D2.1.2/PU/001/24.10.2005.

MAGNET (2005f) Update D3.2.2a candidate air interfaces and enhancements. Deliverable IST-507102 MAGNET/WP3.2/Nokia/D3.2.2b/PU/001/21.12.2005.

MAGNET (2005g) User centric scenarios for PNs of a valid architecture. Deliverable IST-507102 MAGNET/WP1.3/D1.3.1.b/DTU/R/PU/001/1.0.

MAGNET Beyond (2006a) The extended secure architecture – First cycle. Deliverable IST-027396 MAGNET/B/WP4.1/WMC/D4.1.1/R/RE/001/20.12.2006.

MAGNET Beyond (2006b) First solutions for implementation of key management and crypto techniques. Deliverable IST-027396 MAGNET/B/WP4.2/UNIS/D4.2.1/AR/RE/001/1.0/% 20/12/2006.

MAGNET Beyond (2007) Specification of PN networking and security components. Deliverable IST-027396 MAGNET/B/WP2.3/DUT/D2.3.1/PU/001/12.01.2007.

MAGNET Beyond (2008a) Final PN key management solution and cryptographic techniques. Deliverable IST-027396 MAGNET/B/WP4.2/UNIS/D4.2.2/PU/001/1.0/25.06.2008.

MAGNET Beyond (2008b) PN secure networking frameworks, solutions and performance. Deliverable IST-027396 MAGNET/B/WP2.3/DUT/D2.3.2/PU/001/24.09.2008.

MAGNET Beyond (2008c) Usability testing of pilot services. Deliverable IST-027396 MAGNET/B/WP1.4/DTU/D1.4.3/PU/001/30.06.2008.

Maltz, D. and Bhagwat, P. (1998) Msocks: An architecture for transport layer mobility. 17th Annual Joint Conference of the IEEE Computer and Communications Societies (INFOCOM'98), San Francisco, CA, USA.

Maniatis, P., Roussopoulos, M., Swierk, E., Lai, K., Appenzeller, G., Zhao, X. and Baker, M. (1999) The Mobile People Architecture. *ACM SIGMOBILE Mobile Computing and Communications Review (MC2R)*, **3**(3), 36–42.

McDermott-Wells, P. (2004) Bluetooth scatternet models. *IEEE Potentials*, **23**(5), 36–39.

Methley, S. (2009) *Essentials of Wireless Mesh Networking*. Cambridge University Press, Cambridge.

Mirzadeh, S., Tafazolli, R., Armknecht, F., Pallares, J.J. and Afifi, H. (2008a) CPFP: An efficient key management scheme for large scale personal networks. Third International Symposium on Wireless Pervasive Computing (ISWPC'08), Santorini, Greece.

Mirzadeh, S., Tafazolli, R., Pallares, J.J., Armknecht, F. and Afifi, H. (2008b) CPFP: An efficient key management scheme for large scale personal networks. International Symposium on Wireless Pervasive Computing 2008 (ISWPC'08), Santorini, Italy.

Mockapetris, P. (1987) Domain names – Concepts and facilities. IETF RFC 1034, Std. 13.

Montenegro, G.E. (2001) Reverse tunneling for mobile IP, revised. IETF RFC 3024.

Moore, G.E. (1965) Cramming more components onto integrated circuits. *Electronics Magazine*.

Moses, T. (2005) Extensible access control markup language (XACML) version 2.0. OASIS standard. http://docs.oasis-open.org/xacml/2.0/access_control-xacml-2.0-core-specos.pdf.

Moskowitz, R., Nikander, P., Jokela, P. and Henderson, T.R. (2008) Host identity protocol. IETF RFC 5201.

Muñoz, L., Sanchez, L., Lanza, J., Alutoin, M., Ahola, K., Zeghlache, D., Genet, M.G., Hoebeke, J., Moerman, I., Olsen, R.L., Ghader, M., Petrova, M. and Jacobsson, M. (2005) A proposal for self-organizing personal networks. 15th Wireless World Research Forum (WWRF) Meeting, Paris, France.

Murthy, S. and Garcia-Luna-Aceves, J.J. (1996) An efficient routing protocol for wireless networks. *Mobile Networks and Applications*, **1**(2), 183–197.

Narten, T., Nordmark, E., Simpson, W.A. and Soliman, H. (2007) Neighbor discovery for IP version 6 (IPv6). IETF RFC 4861.

Ng, C.W., Ernst, T., Paik, E.K. and Bagnulo, M. (2007a) Analysis of multihoming in network mobility support. IETF RFC 4980.

Ng, C.W., Zhao, F., Watari, M. and Thubert, P. (2007b) Network mobility route optimization solution space analysis. IETF RFC 4889.

Niemegeers, I.G. and Heemstra de Groot, S.M. (2003) Research issues in ad-hoc distributed personal networking. *Wireless Personal Communications: An International Journal*, **26**(2–3), 149–167.

Niemegeers, I.G. and Heemstra de Groot, S.M. (2005) FEDNETS: Context-aware ad-hoc network federations. *Wireless Personal Communications: An International Journal*, **33**(3–4), 305–318.

Nikander, P., Henderson, T.R., Vogt, C. and Arkko, J. (2008) End-host mobility and multihoming with the host identity protocol. IETF RFC 5206.

Norman, D.A. (1988) *The Psychology of Everyday Things*. Basic Books, New York.

Obraczka, K., Viswanath, K. and Tsudik, G. (2001) Flooding for reliable multicast in multi-hop ad hoc networks. *Wireless Networks*, **7**(6), 627–634.

Olsen, R.L. (2008) Enhancement of wide-area service discovery using dynamic context information. PhD thesis Aalborg University, Denmark.

PACWOMAN (2002) System requirements and analysis. Deliverable D2.1, IST-2001-34157 PACWOMAN.

Perkins, C.E. (2001) *Ad Hoc Networking*. Addison Wesley, Boston.

Perkins, C.E. (2002) IP mobility support for IPv4. IETF RFC 3344.

Perkins, C.E. and Bhagwat, P. (1994) Highly dynamic destination-sequenced distance-vector routing (DSDV) for mobile computers. *ACM SIGCOMM Computer Communication Review*, **24**(4), 234–244.

Perkins, C.E. and Royer, E.M. (1999) Ad hoc on-demand distance vector (AODV) routing. Second IEEE Workshop on Mobile Computing Systems and Applications (WMCSA'99), New Orleans, LA, USA.

Perkins, C.E., Belding-Royer, E.M. and Das, S.R. (2003) Ad hoc on-demand distance vector (AODV) routing. IETF RFC 3561.

PNP2008 (2006) Architecture of PNs. Deliverable Freeband/PNP2008/D1.7v1.0.

PNP2008 (2008a) Detailed network-level functionality. Deliverable Freeband/PNP2008/DA.2.1v1.0.

PNP2008 (2008b) Detailed PN management functionality. Deliverable Freeband/PNP2008/DA.2.3v1.0.

PNP2008 (2008c) Detailed service and applications functionality. Deliverable Freeband/PNP2008/DA.2.2v1.0.

PNP2008 (2008d) Federations of Personal Networks. Deliverable Freeband/PNP2008/DA.2.5v1.0.

PNP2008 (2008e) PN architectures – Final version. Deliverable Freeband/PNP2008/DA.1.3v1.0.

PNP2008 (2008f) Updated technical description of PN demonstrator. Deliverable Freeband/PNP2008/DB.2.1v1.0.

Pollin, S., Ergen, M., Timmers, M., Dejonghe, A., van der Perre, L., Catthoor, F., Moerman, I. and Bahai, A. (2006) Distributed cognitive coexistence of 802.15.4 with 802.11. First International Conference on Cognitive Radio Oriented Wireless Networks and Communications (CrownCom'06), Mykonos, Greece.

Postel, J. (1980) User datagram protocol. IETF RFC 768, Std. 6.

Postel, J. (1981) Internet protocol. IETF RFC 791, Std. 5.

Prasad, R.V., Jacobsson, M., Heemstra de Groot, S.M., Lo, A. and Niemegeers, I.G. (2005) Architectures for intra-personal network communication. Third ACM International Workshop on Wireless Mobile Applications and Services on WLAN Hotspots (WMASH'05), Cologne, Germany.

Rekhter, Y., Moskowitz, R.G., Karrenberg, D., de Groot, G.J. and Lear, E. (1996) Address allocation for private internets. IETF RFC 1918.

Richard, III G.G. (2001) Service and Device Discovery – Protocols and Programming. McGraw-Hill, New York.

Rosenberg, J., Mahy, R. and Matthews, P. (2009) Traversal using relays around NAT (TURN): Relay extensions to session traversal utilities for NAT (STUN). IETF Internet-Draft (Work in Progress), draft-ietf-behave-turn-16.

Rosenberg, J., Mahy, R., Matthews, P. and Wing, D. (2008) Session traversal utilities for NAT (STUN). IETF RFC 5389.

Rowstron, A. and Druschel, P. (2001) Pastry: Scalable, decentralized object location and routing for large-scale peer-to-peer systems. 18th IFIP/ACM International Conference on Distributed Systems Platforms (Middleware'01), Heidelberg, Germany.

Rumney, M. (2009) LTE and the Evolution to 4G. Agilent Technologies, Santa Clara, CA.

Sachs, J. (2003) A generic link layer for future generation wireless networking. IEEE International Conference on Communications (ICC'03), Anchorage, AK, USA.

Sachs, J., Wiemann, H., Magnusson, P., Wallentin, P. and Lundsjö, J. (2004) A generic link layer in a beyond 3G multi-radio access architecture. International Conference on Communications, Circuits and Systems (ICCCAS'04), Chengdu, China.

Sanchez, L., Lanza, J., Muñoz, L. and Vila, J.P. (2005) Enabling secure communications over heterogeneous air interfaces: Building private personal area networks. Eighth International Symposium on Wireless Personal Multimedia Communications (WPMC'05), Aalborg, Denmark.

Sanchez, L., Lanza, J., Olsen, R., Bauer, M. and Genet, M.G. (2006) A generic context management framework for personal networking environments. First International Workshop on Personalized Networks (Pernets'06), San Jose, CA, USA.

Schmidt, M. (2002) Subscriptionless mobile networking: Anonymity and privacy aspects within personal area networks. IEEE Wireless Communications and Networking Conference (WCNC2002), Orlando, FL, USA.

Schulzrinne, H. and Wedlund, E. (2000) Application-layer mobility using SIP. ACM SIGMOBILE Mobile Computing and Communications Review (MC2R), 4(3), 47–57.

Schwiderski-Grosche, S., Tomlinson, A. and Pearce, D.B. (2005) Towards the secure initialisation of a personal distributed environment. Technical Report RHUL-MA-2005-09, Department of Mathematics, Royal Holloway, University of London. http://www.rhul.ac.uk/mathematics/techreports.

Sesia, S., Toufik, I. and Baker, M. (2009) LTE: The UMTS Long Term Evolution – From Theory to Practice. John Wiley & Sons, Ltd, Chichester.

SHAMAN (2002) Final technical report – results, specifications and conclusions. Deliverable D13, IST-2000-25350 SHAMAN.

Snoeren, A.C. and Balakrishnan, H. (2000) An end-to-end approach to host mobility. Sixth Annual International Conference on Mobile Computing and Networking (MobiCom'00), Boston, MA, USA.

Soliman, H., Castelluccia, C., ElMalki, K. and Bellier, L. (2008) Hierarchical mobile IPv6 (HMIPv6) mobility management. IETF RFC 5380.

Srisuresh, P. and Egevang, K.B. (2001) Traditional IP network address translator (Traditional NAT). IETF RFC 3022.

Stajano, F. (2000) The resurrecting duckling – what next? Eighth International Workshop on Security Protocols, Cambridge, UK.

Stajano, F. (2002a) *Security for Ubiquitous Computing*. John Wiley & Sons, Ltd, Chichester.

Stajano, F. (2002b) Security for whom? The shifting security assumptions of pervasive computing. International Symposium on Software Security (ISSS'02), Tokyo, Japan.

Stajano, F. and Anderson, R. (1999) The resurrecting duckling: Security issues for ad-hoc wireless networks. Seventh International Workshop on Security Protocols, Cambridge, UK.

Stevens, W.R. (1994) *TCP/IP Illustrated, Volume 1: The Protocols*. Addison-Wesley, Reading, MA.

Stoica, I., Adkins, D., Zhuang, S., Shenker, S. and Surana, S. (2002) Internet indirection infrastructure. ACM SIGCOMM Conference, Pittsburgh, PA, USA.

Stoica, I., Morris, R., Liben-Nowell, D., Karger, D.R., Kaashoek, M.F., Dabek, F. and Balakrishnan, H. (2003) Chord: A scalable peer-to-peer lookup protocol for Internet applications. *IEEE/ACM Transactions on Networking*, **11**(1), 17–32.

Stojmenović, I. and Wu, J. (2004) Broadcasting and activity scheduling in ad hoc networks. In S. Basagni, M. Conti, S. Giardano and I. Stojmenović (eds) *Mobile Ad Hoc Networking*, pp. 205–230. John Wiley & Sons, Inc., Hoboken, NJ.

Sulaiman, T., Sivarajah, K. and Al-Raweshidy, H.S. (2005) Personal identification (PID) in personal area network (PAN). Wireless Personal Multimedia Communications (WPMC'05), Aalborg, Denmark.

Tafazolli, R. (2004) *Technologies for the Wireless Future: Wireless World Research Forum (WWRF)*. John Wiley & Sons, Ltd, Chichester.

Tafazolli, R. (2006) *Technologies for the Wireless Future: Wireless World Research Forum (WWRF), Volume 2*. John Wiley & Sons. Ltd, Chichester.

Tseng, Y.C., Ni, S.Y., Chen, Y.S. and Sheu, J.P. (2002) The broadcast storm problem in a mobile ad hoc network. *Wireless Networks*, **8**(2/3), 153–167.

Tuexen, M., Xie, Q., Stewart, R., Shore, M., Ong, L., Loughney, J. and Stillman, M. (2002) Requirements for reliable server pooling. IETF RFC 3237.

Vaudenay, S. (2005) Secure communications over insecure channels based on short authenticated strings. Advances in Cryptology – CRYPTO 2005: 25th Annual International Cryptology Conference, Santa Barbara, CA, USA.

Vixie, P., Thomson, S., Rekhter, Y. and Bound, J. (1997) Dynamic updates in the domain name system (DNS UPDATE). IETF RFC 2136.

Vulić, N. (2009) Integration of heterogeneous wireless technologies at the UMTS radio access level. PhD thesis Delft University of Technology, The Netherlands.

W3C (2007) SOAP version 1.2 Part 0: Primer (second edition). W3C Recommendation. http://standards.ieee.org/regauth/oui/tutorials/EUI64.html.

W3C (2008) Extensible markup language (XML) 1.0 (fifth edition). W3C Recommendation. http://www.w3.org/TR/2008/REC-xml-20081126/.

Wakikawa, R., Devarapalli, V., Tsirtsis, G., Ernst, T. and Nagami, K. (2009) Multiple care-of addresses registration. IETF RFC 5648.

Weiser, M. (1991) The computer for the twenty-first century. *Scientific American* pp. 94–104.

Whitten, A. and Tygar, J.D. (1999) Why Johnny can't encrypt: A usability evaluation of PGP 5.0. Eighth USENIX Security Symposium, Washington, DC, USA.

Williams, B. and Camp, T. (2002) Comparison of broadcasting techniques for mobile ad hoc networks. Third ACM International Symposium on Mobile Ad Hoc Networking and Computing (MobiHoc'02), Lausanne, Switzerland.

Wireless Strategic Initiative (2000) *The Book of Visions 2000 – Visions of the Wireless World*. http://www. wireless-world-research.org.

Yang, H., Luo, H., Ye, F., Lu, S. and Zhang, L. (2004) Security in mobile ad hoc networks: Challenges and solutions. *IEEE Wireless Communications*, **11**(1), 38–47.

Yee, K.P. (2002) User interaction design for secure systems. Fourth International Conference on Information and Communications Security (ICICS'02), Singapore.

Yuan, W., Wang, X., Linnartz, J.P. and Niemegeers, I.G. (2010) Experimental validation of a coexistence model of 802.15.4 and 802.11b/g networks. *To appear in the International Journal of Distributed Sensor Networks*.

Zandy, V.C. and Miller, B.P. (2002) Transport layer issues: Reliable network connections. Eighth Annual International Conference on Mobile Computing and Networking (MobiCom'02), Atlanta, GA, USA.

Zeiss, J., Sanchez, L. and Bessler, S. (2007) Policy-driven formation of federations between personal networks. 16th IST Mobile and Wireless Communications Summit, Budapest, Hungary.

Zhang, J. and de la Roche, G. (2010) *Femtocells: Technologies and Deployment*. John Wiley & Sons Ltd, Chichester.

Zhou, J., Jacobsson, M. and Niemegeers, I. (2007a) Cross layer design for enhanced quality routing in personal wireless networking. Second International Workshop on Personalized Networks (Pernets'07), Philadelphia, PA, USA.

Zhou, J., Jacobsson, M. and Niemegeers, I.G. (2007b) Cross layer design for enhanced quality personal wireless networking. Sixth Annual Mediterranean Ad Hoc Networking Workshop (Med-Hoc Net'07), Corfu, Greece.

Zhou, J., Jacobsson, M., Onur, E. and Niemegeers, I.G. (2008) Factors that impact link quality estimation in personal networks. Eighth International Symposium On Computer Networks (ISCN'08), Istanbul, Turkey.

Zhou, J., Jacobsson, M., Onur, E. and Niemegeers, I.G. (2009) A novel link quality assessment method for mobile multi-rate multi-hop wireless networks. Sixth Annual IEEE Consumer Communications & Networking Conference (CCNC'09), Las Vegas, NV, USA.

Zhuang, S.Q., Lai, K., Stoica, I., Katz, R.H. and Shenker, S. (2003) Host mobility using an Internet indirection infrastructure. First International Conference on Mobile Systems, Applications, and Services (ACM/USENIX Mobisys), San Francisco, CA, USA.

Zimmermann, P.R. (1995) *The Official PGP User's Guide*. MIT Press, Cambridge, MA.

Related Websites

http://www.3com.com/products/en_US/detail.jsp?tab=features&pathtype=purchase&sku
=3CRXJK10075

> 3Com OfficeConnect Wireless 108 Mbps 11 g XJACK PC Card, Product Specification, accessed in July 2009.

http://www.3gpp.org/

> 3rd Generation Partnership Project (3GPP), accessed in November 2009.

http://www.ambient-networks.org/

> Ambient Networks (AN), accessed in July 2009.

http://awareness.freeband.nl/

> Freeband Awareness, accessed in November 2009.

http://www.bluetooth.com/

> Bluetooth SIG, accessed in October 2009.

http://www.bluetoothtracking.org/

> Bluetooth Tracking, accessed in July 2009.

https://www.cia.gov/library/publications/the-world-factbook/geos/xx.html

> Central Intelligence Agency (CIA), *The World Factbook*, accessed in December 2009.

http://www.ecma-international.org/memento/TC32-PNF-M.htm

> Ecma TC32 – Editing Group on Personal Networks and their Federations, accessed in January 2010.

http://www.enterprise-communications.siemens.com/Open%20Communications/Our%20
Vision-LifeWorks.aspx

> The Siemens LifeWorks Concept, accessed in July 2009.

http://eulersharp.sourceforge.net/

> Euler Proof Mechanism, accessed in October 2009.

Personal Networks: Wireless Networking for Personal Devices   Martin Jacobsson, Ignas Niemegeers and Sonia Heemstra de Groot
© 2010 John Wiley & Sons, Ltd

http://www.ieee802.org/15/

   IEEE 802.15 Working Group for WPAN, accessed in November 2009.

http://www.ieee802.org/21/

   IEEE 802.21, accessed in November 2009.

http://www.ietf.org/html.charters/manet-charter.html

   The Mobile Ad-hoc Networks (MANET) Charter, accessed in July 2009.

http://www.imec.be/pacwoman/Welcome.shtml

   IST PACWOMAN – Power Aware Communications for Wireless Optimised Personal
   Area Networks, accessed in July 2009.

http://www.internetworldstats.com/

   Miniwatts Marketing Group, accessed in April 2008.

http://www.irda.org/

   Infrared Data Association (IrDA), accessed in July 2009.

http://www.isi.edu/nsnam/ns/

   Network Simulator 2, accessed in July 2009.

http://www.kernel.org/

   Linux kernel, accessed in June 2009.

http://www.madwifi.org/

   Madwifi driver, accessed in July 2009.

http://magnet.aau.dk/

   IST MAGNET Beyond – My Personal Adaptive Global Net, accessed in July 2009.

http://mcrypt.hellug.gr/lib/index.html

   Libmcrypt, accessed in July 2009.

http://www.mobilevce.com/

   Mobile Virtual Centre of Excellence, accessed in July 2009.

http://www.olsr.org/

   The olsr.org OLSR daemon, accessed in July 2009.

http://www.openslp.org/

   OpenSLP, accessed in July 2009.

http://www.opensolaris.org/

   OpenSolaris, accessed in July 2009.

http://www.openvpn.net/

    OpenVPN, accessed in July 2009.

http://pnp2008.freeband.nl/

    Freeband Personal Network Pilot 2008 (PNP2008), accessed in July 2009.

http://www.pucc.jp/

    P2P Universal Computing Consortium (PUCC), accessed in July 2009.

http://www.python.org/

    Python Programming Language, accessed in July 2009.

http://qos4pn.irctr.tudelft.nl/

    IOP GenCom QoS for Personal Networks at Home, accessed in July 2009.

http://www.skype.com/

    Skype, accessed in July 2009.

http://vtun.sourceforge.net/tun/

    Ethertap, accessed in June 2009.

http://www.tinyos.net/

    TinyOS, accessed in July 2009.

http://www.ubuntu.com/

    Ubuntu, accessed in July 2009.

http://www.wimaxforum.org

    WiMAX Forum, accessed in November 2009.

http://www.wimedia.org/

    WiMedia Alliance, accessed in November 2009.

http://www.zigbee.org/

    ZigBee Alliance, accessed in November 2009.

Index

access control, 122, *see also* PN
    federation (PNF), access control
ad hoc networking, *see* mobile ad hoc
    network (MANET)
addressing, 67, 81, 102
ambient Networks (ANs), 29, 35
Ananas, 63
anonymity, 138–40
application, 112, 188
application layer mobility, 80, 108
Application programming interface (API),
    111, 116–17
authentication, 122

broadcasting, *see* cluster, broadcasting
    and inter-cluster communication,
    broadcasting

care-of address (CoA), 75, 86, 107
certificate revocation list (CRL), 133–4,
    158
certificate server (CS), 137
certification authority (CA), 158, 173, 180
Certified PN Formation Protocol (CPFP),
    132–3, 137, 178
client, 188
cluster, 41, 46–8
    alternatives, 61
    broadcasting, 73–4, 165–6
    context management, 127, 180
    definition, 59–61, 188
    formation and maintenance, 64, 66–8,
        161–5, 174

requirements, 61–2
routing, 67–9, 72–3, 166, 174, 184
scalability, 62
secure communication, 134–7
service discovery, 123–4
cognition, 12, 18–19
communication domain, 40, 187
communication interface, 39, 187
Contact Networking, 109
content, 18, 119–20, 175, 189
context
    access to, 117
    awareness, 12, 43–4, 115
    definition, 114, 150–1, 187
    management, 120, 127–8, 179–80
    requirement, 18, 35, 114–15
context management node (CMN), 127,
    180
CoolTown, HP, 34

detecting network access (DNA), 79
device, 39–40, 47, 49, 187
distributed hash table (DHT), 179–80

edge router (ER), 83–5, 177
elliptic curve cryptography (ECC), 130
eviction of personal nodes, 53, 133–4
expected transmission count (ETX), 73,
    166
expected transmission time (ETT), 73

federation agent, 148–9, 189
federation manager, 148–9, 189

Personal Networks: Wireless Networking for Personal Devices   Martin Jacobsson, Ignas Niemegeers and Sonia Heemstra de Groot
© 2010 John Wiley & Sons, Ltd